Biosensors as Analytical Tools for the 21st Century

Aleksandr Simonian · Mary Anitha Arugula ·
Paolo Bollella

Biosensors as Analytical Tools for the 21st Century

Principles and Applications

Springer

Aleksandr Simonian
Department of Mechanical Engineering
Auburn University
Auburn, AL, USA

Mary Anitha Arugula
Biotechnology and Energy Materials
CFD Research Corporation
Huntsville, AL, USA

Paolo Bollella
Department of Chemistry
University of Bari Aldo Moro
Bari, Italy

ISBN 978-3-032-01725-3 ISBN 978-3-032-01726-0 (eBook)
https://doi.org/10.1007/978-3-032-01726-0

This Springer imprint is published by the registered company Springer Nature Switzerland AG
The registered company address is: Gewerbestrasse 11, 6330 Cham, Switzerland

Preface

Biosensors are not mere instruments; they serve as conduits through which scientific discovery flows. However, understanding theory alone is insufficient; one must also navigate the practical realm of experimentation. It is in the laboratory where ideas take shape, hypotheses are tested, and innovation succeeds. That's why our book on biosensors goes beyond theory, offering a unique advantage that sets it apart.

Our book is a culmination of over 15 years of teaching the "Biosensing" course at Auburn University, and it bears the fruit of hours spent in laboratories honing experimental protocols and fine-tuning the art of biosensing. What truly sets us apart is our commitment to providing not just theoretical knowledge, but a tangible bridge to the laboratory. Crafted by authorities in biosensing, this book adopts a reader-friendly and approachable writing style. Each chapter is thoughtfully structured to encompass clear learning objectives and concise summaries of the topics under discussion.

Within the pages of this book, readers will discover a treasure trove of real experimental protocols that have been meticulously crafted and rigorously tested. But our commitment doesn't stop there. We go one step further. We provide a companion disk that contains laboratory results and protocols—a valuable resource that transforms theory into practice.

This advantage is more than just convenience; it is an opportunity. An opportunity for researchers and students to delve into the world of biosensors with confidence, armed with the tools and guidance they need to embark on their own experimental journeys. It's a chance to bridge the gap between theory and application, to turn knowledge into action, and to transform curiosity into tangible results.

In an era where biosensors are at the forefront of scientific innovation, our book stands as a testament to our dedication to advancing the field. We believe that by providing both the knowledge and the means to apply it, we empower individuals to not only understand biosensors but to master them. This book, with its multiple real experimental protocols and the companion disk, is our way of opening the doors to discovery and inviting readers to join us on this exciting journey of exploration and innovation.

Auburn, USA — Aleksandr Simonian
Huntsville, USA — Mary Anitha Arugula
Bari, Italy — Paolo Bollella

Contents

About the Authors

Aleksandr Simonian, Ph.D., D.Sc., is a Professor Emeritus at Auburn University and currently serves as the Program Director for the Biosensing Program at the NSF Division of Chemical, Bioengineering, Environmental & Transport Systems (CBET), a position he established in 2009. He has authored over 168 publications in leading scientific journals and has contributed highly cited review articles, book chapters, and patents in the fields of biosensors and functional biomaterials. His work has garnered over 5,950 citations, with an h-index of 41 and an i10-index of 63 (the number of publications with more than 10 citations). Professor Simonian earned his M.S. in Physics from Yerevan State University, his Ph.D. in Molecular Biophysics from the National Academy of Sciences of Armenia, and a D.Sc. in Bioengineering from the Moscow Institute of Applied Biochemistry, completing these degrees in 1969, 1973, and 1993, respectively. In 2011, he was elected as a Foreign Member of the Armenian National Academy.

Mary Anitha Arugula, Ph.D., is a Principal Scientist at CFD Research Corporation (Huntsville, Alabama) with over 12 years of research experience in developing biochemical sensors, including enzyme-based systems for information processing, biomedical and environmental applications. She received her Ph.D. in Environmental Science and Engineering—Health Sciences) from Clarkson University and served as a postdoctoral research associate at the University of Wisconsin-Milwaukee (Dr. Zhen He), University of Utah (Dr. Shelley D. Minteer) and Auburn University (Dr. Aleksandr Simonian). Her postdoctoral research focused on biosensors, biocomputing bioanalytical chemistry, electro-biochemistry, biofuel cells, enzyme extraction, and nano-biosensing.

At CFD, she worked as a Principal Investigator and Co-Investigator for various DoD and NIH funded projects that focused on biofuels (biobattery), point-of-care iron deficiency electrochemical sensors, microfluidic impedance sensors for biofilm establishment and phage detection, and multiomics (miRNA's) point-of-care impedance sensors. She has over 60 publications in leading scientific journals and has contributed review articles, book chapters and patents in the field of biosensors and biofuel cells. She has over 1300 citations, h-index of 19 and i10-index of 22.

Paolo Bollella, Ph.D., is a Tenure Track Assistant Professor at the University of Bari Aldo Moro (Italy), specializing in advanced enzyme-based and edible biosensors. He earned his B.Sc. (2012) and M.Sc. (2014) in Analytical Chemistry and Ph.D. in Pharmaceutical Sciences (2017) from Sapienza University of Rome. He held post-doctoral positions at Åbo Akademi University (Finland) and Clarkson University (USA), where he was promoted to Research Assistant Professor in 2019. Author of 113 peer-reviewed papers (H-index = 35), he has presented over 100 oral contributions and is among the world's top 2% of scientists (Stanford ranking). He serves on the Advisory Board of DirectSens GmbH and is PI or co-PI of national and international projects including SPARKS, REFIN, and RS. He has received several awards, including the 2023 Early Career Analytical Electrochemistry Prize from the International Society of Electrochemistry—Division 1 and the 2022 Young Researcher Award from the Italian Chemical Society.

Chapter 1
Introduction

The evolution of biosensors stems from the broader field of chemical sensors, which are designed to convert chemical information—such as the concentration of a specific component or the presence of a chemical species—into an analytically useful signal. Among chemical sensors, biosensors represent a distinct category that incorporates a biological recognition element tightly coupled with a transducer system. This combination enables specific, sensitive, and often real-time detection of analytes across diverse fields, including medical diagnostics, environmental monitoring, food safety, and biotechnology.

The process of sensing is a specific form of information transfer, and any transmission of information requires the transmission of energy [1].

Generally, the term "sensor" refers to an analytical device designed to detect specific chemicals (analytes) in a given environment, often within a liquid sample.

1.1 Chemical Sensors

Sensors essentially consist of a chemically sensitive interface (sensitive material) and a transducer capable of converting a chemical input (such as gas concentration or molecule/ion concentration) into an output, typically an electrical signal, through conditioning and/or signal processing electronics. The input signal conveys information about chemical entities, such as concentration and identity of unknown species in the gaseous or liquid phase. Traditionally, quantitative analysis of individual components within complex chemical and environmental samples has required sophisticated techniques that involve bulky laboratory procedures (e.g., gas/liquid chromatography, mass spectrometry, etc.). Sensors can respond to various stimuli, including mechanical, electric, or optical signals.

To understand how they work, consider natural sensors found in living organisms. For instance, the human tongue serves as a natural analytical device, detecting tastes

A. Simonian et al., *Biosensors as Analytical Tools for the 21st Century*,
https://doi.org/10.1007/978-3-032-01726-0_1

in liquids and solids through ion/transport exchange mechanisms. Moreover, it can be viewed as an array of sensors functioning in its natural environment. Man-made devices designed to produce and transmit information in the form of specific signals are termed "chemical sensors."

In chemical sensors, recognition is based on the interaction between the recognition probe and the analyte of interest, with recognition mechanisms of varying complexity. The information generated during the recognition event is transformed into an electrical signal through signal transducers. What is their purpose? Sensors determine the presence, concentration, or quantity of specific chemical substances (analytes) that are detectable.

Chemical sensors are devices that provide real-time information about the chemical composition of the target analyte. They consist of a chemically sensitive layer linked to a transducer capable of converting the recognition event into a signal, which can then be translated into a digital electronic output.

Chemical sensors typically consist of a suitable chemically sensitive material interfaced with a transducer, as illustrated in Fig. 1.1. Transduction involves converting the input event into an output electrical signal (analog voltage or current, digital voltage) that is correlated with the quantity of the analyte of interest responsible for triggering it. The output electrical signal then goes through the necessary conditioning, processing, and storage for subsequent analysis. These output signals can manifest in various forms, including optical, mass-related, thermal, or magnetic measurements.

Chemical sensors find applications in various fields, including manufacturing process monitoring, environmental sensing, health monitoring, and homeland security, among others. Direct-reading selective sensors, such as electrochemical sensors, are capable of detecting species in either gas or liquid phases. These sensors achieve molecular selectivity through interactions at the sensor-sample interface. Selectivity is influenced by the analyte's size, shape, or dipolar properties, which are recognized by molecular films, phases, or "receptor sites," resulting in selective binding, absorption, or permeation of the analyte. In subsequent chapters, we will explore how selectivity in direct-reading chemical sensors can be enhanced through the development of analyte-specific films, membranes, and coatings. Miniaturization of these

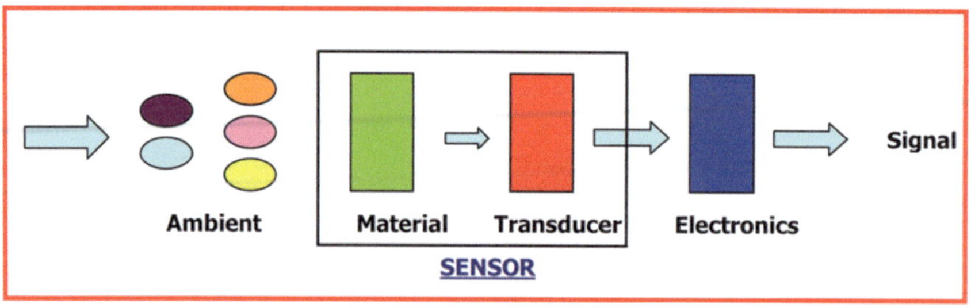

Fig. 1.1 Schematic of a chemical sensor featuring sensitive material, transducer, and interface electronics

sensor systems holds the potential for compact, lightweight, and portable monitoring systems. Notably, the glass pH electrode, developed in 1922, was the first chemical sensor and later adapted for use as a portable device. This device is sensitive to changes in hydrogen ion concentration (protons).

1.2 Biosensors

The significance of biosensors or sensing systems was initially acknowledged within the medical community. For instance, Carl Wunderlich's invention of the thermometer in 1871 marked a crucial milestone in sensing temperature changes. Since the late 1800s, research into sensing and detection systems has driven the development of sensitive and specific sensors in analytical technology. Among the various sensing systems, biosensors stand out as an appealing technology due to their potential for integration with smartphones, paving the way for lab-on-chip and continuous monitoring advancements.

So, what is a "Biosensor"?

A biosensor is a distinct type of chemical sensor capable of detecting biological species based on the presence of a unique biorecognition element that specifically interacts with the target. These biological elements are specifically coupled (immobilized) with a transducing device to convey information in the form of an output signal.

1.2.1 The Discovery of "Biosensor"

The first instance of protein immobilization was reported by Nelson and Griffin in 1916 [2]. They adsorbed the enzyme *invertase* onto activated charcoal and gelatinous aluminum hydroxide without observing any change in enzyme activity.

In 1897, Heinrich Danneel and Walther Nernst made the initial breakthrough in understanding the electrochemical reduction of oxygen. This discovery laid the foundation for further developments in electrochemical methods. Jaroslav Heyrovsky's invention of polarography using the dropping mercury electrode in 1922 marked a significant milestone. It enabled the first measurements of oxygen tension values in plasma and blood in the 1940s [3]. During the same period, Brink, Davies, and Bronk implanted platinum electrodes into tissue to study oxygen supply [4]. However, they focused solely on the behavior of these bare electrodes immersed in blood, overlooking the phenomenon of "biofouling" (later identified as the undesired accumulation of other biological entities on the electrode surface) that would eventually lead to electrode inactivation [5].

In response to these challenges, Leland Clark proposed a solution. He conceived the idea of covering a platinum cathode in glass with cellophane; later, cellophane was replaced with silastic or polyethylene membranes. In 1954, Clark introduced the

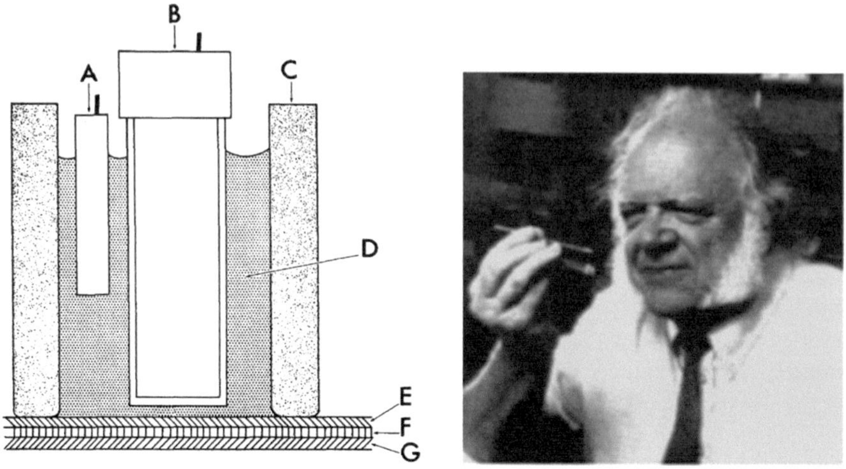

Fig. 1.2 (Left) Clark's first enzyme electrode design. A—reference electrode, B—sensing electrode, C—chamber, D—electrolyte, E and G—membranes and F—enzymes glucose oxidase and peroxidase layer. (Right) Dr. Leland Clark—Inventor of Oxygen Electrode and Father of Biosensors

practical chemical sensor known as the oxygen electrode, as depicted in Fig. 1.2 [6, 7].

In a groundbreaking development in 1962, Clark introduced the concept of biosensors (referred to at that time as an enzyme electrode when he invented the glucose electrode) [8]. His work was highlighted in a landmark address at a New York Academy of Sciences symposium on the evolution of electrochemical sensors, including pH, polarographic, potentiometric, or conductometric sensors, by incorporating enzyme transducers within membrane-enclosed configurations.

The concept of the glucose biosensor was initially demonstrated by entrapping glucose oxidase at Clark's oxygen electrode using a dialysis membrane. Glucose oxidase is an enzyme that catalyzes the oxidation of glucose while consuming oxygen. The decrease in the measured concentration of oxygen is directly proportional to the concentration of glucose, particularly within a specific range of glucose concentrations. The term "enzyme electrode" was first coined by Clark and Lyons to describe these electrodes capable of measuring oxygen levels in blood. Later, a similar device with glucose oxidase entrapped in a polyacrylamide gel was described by Updike and Hicks in 1967 [9].

Clark's device operates on the principle of decreasing oxygen concentration at the electrode due to the consumption of oxygen through glucose oxidation in the biosensing layer, which comprises enzymes, notably glucose oxidase and peroxidase.

In 1962, Guilbault and his coworkers implemented the first analytical application of immobilized enzymes for the early detection of nerve agents [10]. Guilbault's device consisted of two platinum grid electrodes with a porous sorbent tablet containing the immobilized enzyme choline esterase, used for detecting biological targets in chemical warfare prevention [11]. In 1969, Guilbault and Montalvo invented the first potentiometric biosensor for the detection of urea, utilizing urease immobilized on an ammonia electrode [12].

In 1975, several inventions marked significant progress: the development of a fiber-optic sensor with an immobilized indicator to measure carbon dioxide or oxygen, the introduction of the first commercial biosensor (Yellow Springs Instruments glucose biosensor) based on amperometric detection of hydrogen peroxide, and the creation of the first immunosensor on a potentiometric transducer by Janata [13, 14].

In 1963, Garry A. Rechnitz and S. Katz published one of the earliest papers in the field of biosensors, discussing the direct potentiometric determination of urea following urease hydrolysis [15, 16]. These devices were initially referred to as "enzyme electrodes" or "biocatalytic membrane electrodes," as the term "biosensor" had not yet been coined. In 1964, enzymes were first used as fuel catalysts by Yahiro et al. in a glucose/O_2 biofuel cell [17].

The term "Biosensor" was first coined by Karl Cammann in 1977 [18].

Clark's discovery of the glucose electrode sparked the biosensor revolution in biology and chemistry. This discovery not only clarified the basic concept but also demonstrated its potential for a wide range of electrochemical and optical transducers, in combination with antibodies and microbes, to create real-time bioanalytical instruments. Subsequent advancements, driven by the development of immunosensors based on bioaffinity, laid the foundation for the second major evolutionary line in biosensor technology. During this time, biosensor research accelerated significantly, encompassing immobilized enzymes, fiber-optic systems with antibodies, and membrane receptors on piezoelectric crystals. Figure 1.1 illustrates a simplified view of molecular recognition elements and transducers [18].

The major discoveries of this bioanalytical revolution and the focus on current advancements are summarized in Table 1.1. Notably, the period from 1980 to 1993 witnessed significant breakthroughs in biosensor research, sparking a surge in interest and investment in the manufacturing of enzyme-based systems coupled with mediated electrochemistry. This wave of innovation led to the development of self-testing devices for various diseases, with a particular emphasis on diabetes [19].

The momentum continues to this day, with home-use and convenient blood glucose testing equipment and test strips [20] boasting an estimated turnover exceeding US $26.8 billion in 2022 and is expected to grow at a compound annual growth rate (CAGR) of 8.0% from 2023 to 2030 [21]. This substantial market success has further incentivized research efforts aimed at creating new sensing devices. Researchers have harnessed this wealth of knowledge to develop technologies for detecting a wide range of biochemical analytes, contributing to disease prevention, wellness maintenance, and an improved quality of life [21]. Many of these technologies have swiftly entered the market. Notable examples include cholesterol self-testing devices [22], which arguably achieved a higher market penetration rate than glucose self-testing devices, as well as monitoring health indicators beyond the physician's office.

This technology boom has spurred interest in and expanded the capabilities of environmental detection for toxic materials in natural resources. Biosensors have

Table 1.1 Crucial developments in the history of biosensors

Date	Person	Enzyme	Electrode	Analyte	Transducer
1962	Leland Clark	Glucose oxidase	Glucose	Glucose	Amperometric
1963	Guilbault	Choline esterase	Pt	Nerve agent	Amperometric
1964	Yahiro	Glucose oxidase		Glucose	Biofuel cell
1969	Guilbault and Montalvo	Urease	Ammonia	Urea	Potentiometric
1975	Racinee and Mindt	Cytochrome b_2	Platinum	Lactate	Amperometric
1975	Lubbers and Opitz	Alcohol oxidase	Oxygen sensor	Alcohol	Optical
1982	Schultz	–	–	Glucose	Fluorescence
		Antibody	*Surface*	*Antigen*	
1972	Shons		Gold		Piezoelectric
1975	Janata	Antibodies			Potentiometric
1979	Aizawa	Antibodies			Amperometric
1983	Liedberg, Nylander, and Lundström	Human IgG	600 A silver film	Anti-human IgG	Surface-plasmon resonance
		Microbes	*Electrode*		
1975	Divies	*Acetobacter xylinum* (bacteria)	Oxygen	Ethanol	Amperometric
1977	Karube	Soil micro-organisms	Oxygen	Glucose-glutamic acid	Amperometric
1980	Kulys and Kadziaskrene	*Hansenula anomala* (fungi)	Oxygen	Lactate, pyruvate, glucose	Amperometric

found applications in security and surveillance, enabling the identification of chemical warfare agents [23, 24]. They made significant contributions to fields such as food quality and safety control and biomedicine. In diagnostics, electrochemistry played major part in biosensor development, while optical transducers have primarily been confined to research and development. Mass, thermal, and magnetic transducers have played notable roles in biosensor evolution, with the discovery of the piezoelectric quartz crystal microbalance (a mass sensing device) immunosensor by Shons in 1972 [25, 26]. Despite these devices are noteworthy, they have yet to demonstrate significant practical impact to date. Table 1.2 shows the critical inventions that are the beginning of the evolution of more complex and more sensitive biosensors [27].

Table 1.2 Critical inventions in the history of biosensors

Critical inventions	Inventor/company, year
Glass pH electrode	Hughes, 1922
Oxygen electrode	Clark, 1954
Gas-phase biosensor based on QCM	King, 1964
Ion field-effect transistor	Bergveld, 1970
Commercial glucose biosensor	Yellow Springs Instrument, 1975
Optode/fiber-optic sensor	Lubbers, 1975
Electrochemical biosensor introduced in bedside artificial pancreas	Miles, 1976
Fiber-optic pH sensor	Peterson, 1980
Printed biosensors	MediSense, 1980s
Ferrocene mediated biosensor	Hill, 1981
Implantable—needle-type subcutaneous electrode	Shirchiri,1982
Flow-injection immunoassay	Uditha de Alvis and Wilson, 1985
Blood glucose biosensor	MediSense Exactech, 1987
SPR-based biosensor	Pharma Biacore, 1990
Emerging technology-SELEX	Larry Gold and Jack Szostak's, 1990
Handheld blood biosensor	I-STAT, 1992
Launching Gluco card	1998

1.2.2 Development of Biosensor Technology

The expansion of biosensor technology has been remarkable, as exemplified by the launch of the journal "Biosensors and Bioelectronics" by Elsevier in 1985 by Professor A. P. F. Turner [28]. During the 1990s, research and development in bioanalytical probing focused on enhancing the stability and sensitivity of biosensing devices [29, 30]. However, much of the research during this period was significantly constrained by the lack of efficient preparation methods, including separation, isolation, purification, and the viability of microorganisms or living cells for new enzymes or enzyme sources.

The trends and developments from 1980 to 2000 prominently featured bioaffinity sensors. Beyond enzymes, antibodies, and microbes, DNA probes and nucleic acid biosensors gained increasing importance, particularly DNA hybridization based on sequence-specific information, applied in clinical, environmental, and forensic investigations [31, 32]. Another notable development centered on high-throughput screening, demanding miniaturized, rapid, and cost-effective detection methods with label-free transduction. Bioengineered proteins attracted particular attention for detecting environmental heavy metals at low concentrations. To achieve this, self-assembled monolayers of thiol on gold electrodes were fabricated, to which proteins were bound using a cross-linker, effectively acting as "glue," for the selective detection of heavy metals [33–35]. Moreover, a new trend emerged involving biomimetic receptors, synthetic or artificial receptors, instead of enzymes, antibodies, or biological moieties, offering inherent advantages. Molecular imprinting, a technique for inducing affinity sites toward specific compounds, was frequently used in conjunction with bioaffinity sensors [36, 37]. Layer-by-layer technique emerged due to its simplicity and versatility, has emerged as one of the most popular approaches for the fabrication of biofunctionalized multilayers, based on the alternating assembly of oppositely charged layers. It has been demonstrated to be an effective approach for the attachment of a variety of enzymes onto carbon nanotubes (CNTs) and self-assembly of organized CNT-biopolymer/CNT-enzyme multilayered biosensors [38–40].

Meanwhile, enzyme-based glucose sensors continued to dominate the market, with ongoing efforts focused on implantable sensors. Zoll's initial report on electrical heart stimulation dates back to 1952, and the first commercially available wireless blood pressure measurement system was introduced by CardioMEMS in 2010 [41]. Researchers made considerable efforts to enhance patients' quality of life through various medical devices, including implantable cardiac defibrillators, cochlear implants, implanted bladder stimulators, and implantable wireless pressure sensors. Enzyme-based glucose implants demonstrated remarkable stability, often lasting three months or more. Lactose sensors, designed for monitoring body fluid build-up during muscle training, were launched with new immobilization methods enabling direct electron transfer. Investigations into innovative concepts at the electrode interface were a new trend, led by researchers like Willner and his colleagues. More advanced biosensors using whole cells with specific metabolic functions were developed as an alternative. A novel approach involved the implementation of fiber-optic and cell-implantable nanosensors for testing pH, calcium, nitric oxide, and glucose in rat embryos and single mammalian cells. Optical imaging fibers, ion-selective field-effect transistors (FETs), and oxygen sensors played pivotal roles in understanding chemical analysis at the single-cell level, cellular functions, cell-cell communications, and neural networks [42, 43]. Progress toward miniaturization of analytical instruments using microfabrication and nanofabrication techniques represented a significant departure from "traditional" biosensors. This miniaturization offered the potential to scale down setups to molecular dimensions. Consequently, well-established microsystem technologies were adapted, signal processing steps were integrated onto chips, and arrays were built for complex pattern recognition analysis.

In the past decade, a variety of transducer principles have been implemented, shifting the focus from "bio" to "sensing" scenarios. Surface modification of biosensing interfaces using stabilizers, ionophores, additives, cross-linkers (for protein stability), specific functional groups for enzyme labeling, and solid supports has gained attention. A major leap forward was made with the introduction of the "microcontact printing" technique, enabling the patterned immobilization of proteins on solid substrates with an exceptional 99% efficiency [42–44]. As these new trends progressed, it was anticipated that classical sensor technologies would converge with emerging technologies, such as scanning probe microscopy (atomic force microscopy—AFM), to study and characterize antigen-antibody binding and detect single antigen molecules.

1.2.3 The Biosensor Revolution

Glucose sensors, predominantly utilizing glucose oxidase (GOx), have historically dominated the field of biosensors. However, recent years have witnessed the emergence of engineered proteins based on pyrroloquinoline quinone dehydrogenase and other catalysts and affinity elements, surpassing traditional GOx-based biosensors. These advancements involve the creation of hybrid enzymes through genetic manipulation of catalytic proteins. Allosteric enzymes, incorporated through modular engineering, serve as sensing elements, strategically positioned to enable the enzyme to catalytically respond to new effectors. A notable example of such an engineered allosteric biosensor is *Escherichia coli* β-galactosidase, composed of four identical subunits, which serves as a reporter, marker, and partner in fusion proteins, facilitating structurally responsible architecture for potential biosensing applications. These hybrid enzymes, when combined with redox centers, exhibit the highest affinity and specificity toward desired metabolites.

While enzymes have had a significant impact, particularly in the form of enzyme electrodes, DNA molecules have demonstrated their capacity to catalyze a range of reactions and are commonly referred to as DNAzymes. Similar to proteins, DNAzymes can recruit cofactors like metal ions to create metallo-enzymes with high activity for cleaving DNA and RNA, offering great promise for pharmaceutical applications and metal ion sensors. Electrochemical sensors have delivered remarkable results and are widely used in the market. However, field-effect transistors, a form of electrochemical sensing technology that monitors charge effects resulting from biochemical interactions, have shown great promise [45, 46].

Affibodies, smaller than traditional monoclonal and polyclonal antibodies, have been modified to provide increased access to tissues and smaller epitopes than natural antibodies. Synthetic antibodies, commonly referred to as aptamers, represent a significant advancement in the field of antibodies. Initially emerging from laboratory experimentation, synthetic antibodies have now gained prominence as reagents for biological research and medical applications. This is evident from the substantial investments made in companies with the capability to produce these remarkable molecules [47, 48].

Nanocarbons, including carbon nanotubes (CNTs), graphene, and fullerenes, have gained prominence as materials for designing next-generation functional materials used in the quantitative and qualitative detection of chemical and biological entities for medical, environmental, and energy applications. CNTs and graphene, with their high surface area, electrical properties, mechanical strength, and modulation capabilities with analytes, hold great promise for chemical and biosensing applications. While several other nanomaterials exhibit excellent functions, CNTs, originally a technological breakthrough in 1991, have been extensively studied and recognized for their unique physicochemical properties, such as good electrical conductivity, high stability, and strong adsorptive ability. These materials continue to be vital components of current and future biosensor technologies [48, 49].

Emerging hybrid nanostructures, modified with sensing elements, offer a compelling combination of intrinsic physical and chemical properties suitable for various applications, including mimetic biominerals, biomedicine, and biosensors. Historically, sensors were constructed using pristine and functionalized CNTs. However, selectivity can be an issue due to the limited interaction between the analyte and CNTs. Sensing elements can range from synthetic molecules to biomolecules. Integrating CNTs with various biological molecules, such as biopolymers, biocatalysts, and biorecognition structures, yields intriguing electrical, mechanical, and optical properties. A wealth of research has been published on the development of biosensors that incorporate biocatalysts/enzymes with CNTs and various nanomaterials, such as magnetic nanoparticles (MNPs) and quantum dots (QDs), for the detection of DNA, RNA, proteins, pesticides, glucose, and other small molecules [50–54]. These materials possess both metallic and semiconducting properties, which not only enhance direct electron transfer but also provide excellent scaffolding structures for enzymes, maintaining their bioactivity. This results in faster response times, signal amplification, higher sensitivity, improved selectivity, and stability. However, the feasibility of CNTs depends on their synthesis and the ability to achieve homogeneous dispersion while retaining structural integrity. Several methods have demonstrated that the presence of oxidants, such as strong acids, can facilitate uniform distribution, create a tight matrix, and enable functionalization with organic groups. The electrocatalytic properties of CNTs arise from their sidewall pi-conjugated structure with a highly hydrophobic surface and the presence of carboxylic groups, allowing for interactions with polyelectrolytes through pi-pi electronic and hydrophobic interactions, thus minimizing aggregation and enhancing the ability to immobilize biomolecules. Examples of polyelectrolytes include aqueous sodium dodecyl sulfate (SDS), Nafion, polyethylene amine (PEI), DNA, among others [38, 39, 54, 55]. Enzyme-based biosensors have been one of the most well-developed and successful areas in nanobiosensing.

1.3 Integration of Biosensor Components

Biosensors employ biological reactions to detect target analytes and consist of a biological component and a transducer. The biological components are commonly referred to as "biocatalysts," sometimes as "biological elements," and in recent literature and textbooks as "molecular recognition elements," "biological recognition elements," or "biological sensing elements." These biological entities possess recognition properties that enable them to bind specifically and selectively to the target analyte. Since a variety of recognition elements exist in cells and molecular systems, they can be classified into two categories.

Table 1.3 Examples of bioreceptor and the analyte

Receptor (R)	Analyte (S)
Antibody	Antigen
Dye	Protein
Lectin	Saccharide
Apoenzyme	Prosthetic group
Receptor	Hormone
DNA	DNA

Bioaffinity: This category of recognition is based on strong attractive bonds between the bioreceptor and the analyte: S + R ↔ SR. Examples are shown in Table 1.3.

The primary components of recognition in biosensors include antibody-antigen and DNA–DNA interactions. *Antibodies* are proteins generated in mammals and higher vertebrates to serve as the first line of defense against invading pathogens. Pathogens and similar substances that stimulate antibody production are referred to as *antigens*. This immune response forms the basis for sensors that utilize these immune interactions, which are known as immunosensors. Antibodies function by tightly binding to specific antigens, forming precise complexes on surfaces. The mechanism of interaction, or the molecular and structural binding events, is primarily governed by the properties of the antibody molecules and transducers, achieving high sensitivity and selectivity, especially in complex matrices. Chapter 3 will provide a detailed focus on immunosensors.

Similarly, interactions between two DNA strands involve base pairing, allowing the structures to fit together like pieces of a jigsaw puzzle. Sensors fabricated with nucleic acids and their derivative sequences are generally referred to as DNA sensors or DNA biosensors, used for the detection of target sequences. DNA probes, short nucleotide sequences, exhibit strong and stable attraction toward DNA fragments specific to certain genes. Biosensors of this kind are termed genosensors.

Biocatalysts: Recognition in this category is based on biochemical reactions involving substrate consumption and product formation. S + R ↔ SR → P + R. Examples are shown in Table 1.4.

The most commonly used biosensors are enzyme-based biosensors, also known as enzyme sensors or enzyme electrodes. Enzymes are proteins that catalyze the conversion of reactants (substrates) into products, often accompanied by a measurable signal. For instance, glucose oxidation catalyzed by *glucose oxidase* (a single

Table 1.4 Examples of bioreceptor and the analyte and product formation

Receptor (R)	Analyte (S)
Enzyme	Substrate
Organelle	Cofactor
Microbe	Inhibitor
Tissue slice	Activator

enzyme) results in the production of gluconic acid and hydrogen peroxide (H_2O_2). Another example is the hydrolysis of urea, which is catalyzed by *Urease*.

$$\text{Glucose} + O_2 \xrightarrow{\text{Glucose oxidase}} \text{Gluconic acid} + H_2O_2$$

$$\text{Urea} \xrightarrow{\text{Urease}} 2NH_3 + CO_2$$

Biological tissues and organelles represent another class of biocatalysts that have been utilized to maintain the desired enzyme in its natural environment and preserve its activity. Tissue materials have garnered particular interest because they can offer sufficiently high specific activity for the development of specific biosensors, especially in cases where the extraction and purification of enzymes have proven challenging or impractical. This approach allows for the construction of biosensors that can function effectively in complex biological environments.

$$\text{Arginine} \xrightarrow{\text{Bovine Liver}} \text{Urea}$$

$$\text{Urea} \xrightarrow{\text{Urease}} 2NH_3 + CO_2$$

Microbial cells, whole cells, or cultivated tissues of organisms have been harnessed for sensing purposes. However, their implementation typically involves immobilized multicellular organisms integrated with an electrode. The development of microbial sensors has demonstrated superior biocatalytic activity, enabling the detection of inert substances such as methane and ammonia through the involvement of reaction sequences and biological pathways. Microbial sensors introduce a novel approach to microbioassays, facilitating the detection of mutagens and the identification of vitamins [56]. These sensors offer a straightforward and cost-effective method for analyzing microbiological samples by detecting saccharides, organic compounds, carbohydrates, and assimilable substances produced by microbial cells.

A transducer plays a crucial role in converting the biological or biochemical response resulting from the interaction with a target analyte into a quantifiable signal. Commonly used transducing elements include electrochemical (electric current), optical (light), and thermal (heat) components.

Electrochemical biosensors are typically associated with biocatalytic recognition elements that either produce or consume electrons. For instance, redox enzymes like glucose oxidase or glucose dehydrogenase are frequently employed. These sensors are often configured in three-electrode cells, consisting of a working electrode (where electron-related reactions occur), a reference electrode (providing a reference potential to bias the working electrode), and a counter/auxiliary electrode (facilitating current flow through the external circuit). The target analyte participates in a reaction on the active electrode surface, which may result in electron transfer across the double layer, generating a current. Alternatively, it can contribute to the double-layer potential, leading to voltage. Analyte concentration can be determined by measuring either the current (which is proportional to the rate of electron flow) at a fixed potential or the potential at zero current, which produces a logarithmic response, as described by the Nernst equation.

Additionally, there is a third type of electrochemical biosensor based on electrochemical impedance spectroscopy. These sensors are primarily associated with bioaffinity recognition elements. Binding events at the electrode surface lead to an increase in steric hindrance or changes in the charge of the double layer, resulting in changes in resistance to charge transfer, which can be quantitatively measured.

Optical and thermal biosensors are versatile and can accommodate both bioaffinity and biocatalytic recognition elements. These sensors can induce changes in optical properties (such as absorbance, fluorescence, surface-plasmon resonance, etc.) and thermal properties (including enthalpy) based on interactions between substrates and enzymes, antigens and antibodies, or DNA strands.

Apart from these three primary groups of transducers, there are other types that have seen limited success due to specific characteristics like instability, low sensitivity, or lack of selectivity. Examples include piezoelectric, gravimetric, and magnetic biosensors. Additionally, field-effect transistors are another group of transducers capable of detecting analytes at the single-molecule level, with the first report in 2018 [57, 58].

Recognition elements and transducers are linked with electronics or signal processors that are responsible for presenting the results in a user-friendly manner. There have been significant efforts, especially from an engineering and design perspective, to achieve effective miniaturization and customization of biosensor devices, transitioning from bench-top to portable systems.

References

1. Fraden, J. Handbook of Modern Sensors. Physics, Designs, and Applications. *Springerlink* (2010) https://doi.org/10.1007/978-1-4419-6466-3.
2. Griffin, J. M. N. and E. G. ADSORPTION OF INVERTASE. *J. Am. Chem. Soc.* **38**, 1109–1115 (1916).
3. Bard, A. J. The rise of voltammetry: From polarography to the scanning electrochemical microscope. *J. Chem. Educ.* **84**, 644–650 (2007).
4. H. Danneel, V. Über den durch diffundierende Gase hervorgerufenen Reststrom. *Zeitschrift Fur Elektrochemie,* 227–231 (1897).
5. Severinghaus JW, A. P. History of blood gas analysis. IV. Leland Clark's oxygen electrode. *J Clin Monit.* **2(2)**, 125–139 (1986).
6. Freeman, A. Electrodes for Blood PO_2 and pCO_2 Determination. *J. Appl. Physiol.* **13**, 515–520 (1958).
7. Clark Jr, LC; Wolf, R; Granger, D; Taylor, Z. Continuous recording of blood oxygen tensions by polarography. *J. Appl. Physiol.* **6**, 189–193 (1953).
8. CLARK, L C Jr, and C. L. Electrode systems for continuous monitoring in cardiovascular surgery. *Ann. N. Y. Acad. Sci.* **102**, 29–45 (1962).
9. Updike, S J, and G. P. H. The enzyme electrode. *Nature* **214**, 986–988 (1967).
10. Kramer, D. N., P. L. Cannon Jr, and G. G. G. Electrochemical Determination of Cholinesterase and Thiocholine Esters. *Anal. Chem.* **34**, 842–845 (1962).

11. Guilbault, G. G., Sadar, M. H., Kuan, S. S., & Casey, D. Enzymatic methods of analysis: Trace analysis of various pesticides with insect cholinesterases. *Anal. Chim. Acta* **52**, 75–82 (1970).
12. Guilbault, George G., and J. G. M. J. Enzyme electrode for the substrate urea. *J. Am. Chem. Soc.* **92**, 2533–2538 (1970).
13. Lübbers, D W, and N. O. "Die pCO_2-/pO_2-Optode: Eine neue pCO_2- bzw. pO_2-Messsonde zur Messung des pCO_2 oder pO_2 von Gasen und Flüssigkeiten" [The pCO_2-/pO_2-optode: a new probe for measurement of pCO_2 or pO in fluids and gases (authors transl)]. *Zeitschrift fur Naturforschung. Sect. C, Biosci.* **30**, 532–533 (1975).
14. Janata, J. Immunoelectrode. *J. Am. Chem. Soc.* **97**, 2914–2916 (1975).
15. Rechnitz, G. A. Cation-sensitive glass electrodes. *J. Chem. Educ.* **41**, 385 (1964).
16. Katz, S. A. Direct Potentiometric Determination of Urease Activity. *Anal. Chem.* **36**, 2500–2501 (1964).
17. Yahiro, A.T. & Lee, S.M. & Kimble, D. O. Bioelectrochemistry: I. Enzyme Utilizing Bio-Fuel Cell Studies. *Biochim. Biophys. Acta - Spec. Sect. Biophys. Subj.* **88**, 375–383 (1964).
18. Taylor, R. F. and Schultz, J. S. Handbook of Chemical and Biological Sensors. Institute of Physics, Philadelphia, in 604 (1996).
19. Anthony P.F. Turner, J. C. P. Diabetes mellitus: biosensors for research and management. *Biosensors,* **1**, (1985).
20. MD., H. The business of self-monitoring of blood glucose: a market profile. *J Diabetes Sci Technol.* **3**, 1219–1223 (2009).
21. Chen C, Zhao XL, Li ZH, Zhu ZG, Qian SH, F. A. Current and Emerging Technology for Continuous Glucose Monitoring. *Sensors (Basel).* **17**, 182 (2017).
22. Taylor JR, L. L. Cholesterol: Point-of-Care Testing. *Ann. Pharmacother.* **38**, 1252–1257 (2004).
23. M., P. Current Trends in the Biosensors for Biological Warfare Agents Assay. *Mater. (Basel).* **12**, 2303 (2019).
24. E. Rainina, A. Simonian, A. Efremenco, S. Varfolomeyev and J. Wild, Biosensors & Bioelectronics, Vol. 11, No 10, pp. 991–1000 (1996).
25. Shons, Alan, Frank Dorman, and J. N. An immunospecific microbalance. *J. Biomed. Mater. Res.* **6**, 565–570 (1972).
26. Lambrechts, M.; Sansen, W. B. M. D. I. of P. P. B. & UK, 1992; ISBN 978-0750301121. Biosensors: microelectrochemical devices. *Bristol; UK; Inst. Phys. Publ.* **ISBN 978-0**, (1992).
27. Trupti R. Das, Santanu Patra, Penny P. Govender, S. K. S. Biosensors: principle, fundamentals history, recent trends and applications, Chapter 1. *Biosens. Emerg. Re-Emerging Infect. Dis. Acad. Press. 2022, Pages 1–18, ISBN 9780323884648.*
28. Turner, A., Karube, I., & Wilson, G. S. Biosensors: Fundamentals and Applications (1st ed.). Retrieved, (1987).
29. Scheller, F., Schubert, F., Biosensors (Techniques and Instrumentation in Analytical Chemistry, Volume 11) 1st Edition, Elsevier Science; 1st edition (1992), ISBN-10: 0444987835.
30. Renneberg, R and Lisdat, F (eds.): Biosensing for the 21st century, Series: Advances in Biochemical Engineering/Biotechnology, Vol. 109, Springer, ISBN: 978-3-540-75200-4.
31. Songjun Li, A. T. Dedication and prominence: 31 years in 'Biosensors and bioelectronics'. *Adv. Mat. Lett.* **2**, 84–89 (2011).
32. Taylor R.F. and Schultz, J. S. Handbook of Chemical and Biological Sensors. *Inst. Phys. Publ. Bristol, UK.* (1996).
33. Christopoulos, T. K. Nucleic acid analysis. *Anal. Chem.* **71**, 425–438 (1999).
34. Palecek E, Fojta M, Tomschik M, W. J. Electrochemical biosensors for DNA hybridization and DNA damage. *Biosens Bioelectron* **13**, 621–628 (1998).
35. Herrera-Chacón, A., Cetó, X. & del Valle, M. Molecularly imprinted polymers - towards electrochemical sensors and electronic tongues. *Anal Bioanal Chem* **413**, 6117–6140 (2021).
36. Willner, I., Katz, E., Riklin, A., & Kasher, R. Mediated electron transfer in gluthathione reductase organized in self-assembled monolayers on gold electrodes. *Journal Am. Chem. Soc.* **114**, 10965–10966. (1992).
37. Vimala Rani Samuel, K. J. R. A review on label free biosensors. *Biosens. Bioelectron. X* **11**, (2022).

38. Arugula, M. A., & Simonian, A. Novel trends in affinity biosensors: current challenges and perspectives. *Meas. Sci. Technol. 25(3), 032001.* **25**, 032001 (2014).
39. Arugula, M., & Simonian, A. Nanocarbon-Based Multi-Functional Biointerfaces: Design and Applications. *ECS J. Solid State Sci. Technol. 5(8), M3045.* **5**, M3045 (2016).
40. Zhang, Y., Arugula, M. A., Wales, M., Wild, J., & Simonian, A. L. A novel layer-by-layer assembled multi-enzyme/CNT biosensor for discriminative detection between organophosphorus and non-organophosphrus pesticides. *Biosens. Bioelectron. 287–295.* **67**, 287–295 (2015).
41. Philip B. Adamson, William T. Abraham, Robert C. Bourge, Lynne W. Stevenson, J. Y. CardioMEMS Heart Sensor Allows Monitoring of Pressures to Improve Outcomes in NYHA Class III Heart Failure Patients (CHAMPION) Trial: Impact of Hemodynamic Guided Care on Patients With Preserved Ejection Fraction. *J. Card. Fail.* **16**, 913 (2010).
42. Andreas Steinegger, Otto S. Wolfbeis, and S. M. B. Optical Sensing and Imaging of pH Values: Spectroscopies, Materials, and Applications. *Chem. Rev. 2020 120 (22), 12357–12489* **120**, 12357–12489 (2020).
43. Saman Sargazi, Iqra Fatima, Maria Hassan Kiani, Vahideh Mohammadzadeh, Rabia Arshad, Muhammad Bilal, Abbas Rahdar, Ana M. Díez-Pascual, R. B. Fluorescent-based nanosensors for selective detection of a wide range of biological macromolecules: A comprehensive review. *Int. J. Biol. Macromol.* **206**, 115–147 (2022).
44. Mrksich, M., & Whitesides, G. M. Patterning self-assembled monolayers using microcontact printing: a new technology for biosensors? *Trends Biotechnol.* **13**, 228–235. (1995).
45. Bernard, A., Renault, J. P., Michel, B., Bosshard, H. R., & Delamarche, E. Microcontact printing of proteins. *Adv. Mater.* **12**, 1067–1070 (2000).
46. Schöning, M. J., & Poghossian, A. Recent advances in biologically sensitive field-effect transistors (BioFETs). *Analyst,* **127**, 1137–1151 (2002).
47. 50. Turner, A. P. Biosensors: sense and sensibility. *Chem. Soc. Rev.* **42**, 3184–3196 (2013).
48. Mascini, M., Palchetti, I., & Tombelli, S. Nucleic acid and peptide aptamers: fundamentals and bioanalytical aspects. *Angew. Chemie Int. Ed. 51(6), 1316–1332.* **51**, 1316–1332 (2012).
49. Alkire, Richard C., Yury Gogotsi, and P. S. Nanostructured materials in electrochemistry. *John Wiley Sons,* (2008).
50. Harris, P. J. Carbon nanotube science: synthesis, properties and applications. *Cambridge Univ. Press.* (2009).
51. Kumar, Challa SSR, E. Nanomaterials for biosensors. *John Wiley Sons,* (2007).
52. Balasubramanian, K., & Burghard, M. Biosensors based on carbon nanotubes. *Anal. Bioanal. Chem.* **385**, 452–468 (2006).
53. Agüí, L., Yáñez-Sedeño, P., & Pingarrón, J. M. Role of carbon nanotubes in electroanalytical chemistry: a review. *Anal. Chim. Acta,* **622**, 11–47 (2008).
54. Lin, Y., Yantasee, W., & Wang, J. Carbon nanotubes (CNTs) for the development of electrochemical biosensors. *Front. Biosci.* **10**, 492–505 (2005).
55. Wang, J. Carbon-nanotube based electrochemical biosensors: A review. *Electroanal. An Int. J. Devoted to Fundam. Pract. Asp. Electroanal.* **17**, 7–14 (2005).
56. Zhang, Y., Arugula, M. A., Kirsch, J. S., Yang, X., Olsen, E., & Simonian, A. L. Layer-by-layer assembled carbon nanotube-acetylcholinesterase/biopolymer renewable interfaces: SPR and electrochemical characterization. *Langmuir* **31**, 1462–1468 (2015).
57. Karube, I., Matsunaga, T., Mitsuda, S., & Suzuki, S. Microbial electrode BOD sensors. *Biotechnol. Bioeng.* **19**, 1535–1547. (1977).
58. Vu C-A, C. W.-Y. Field-Effect Transistor Biosensors for Biomedical Applications: Recent Advances and Future Prospects. *Sensors* **19**, 4214 (2019).

Chapter 2
Biosensors Characteristics: Definitions and Concepts

The performance of a biosensor is largely determined by its transducer. In an ideal system, composed of perfect materials and flawless interfaces, the transducer would produce an absolute and reproducible output in direct response to an applied stimulus. However, practical limitations—such as material imperfections, noise, interference, and environmental variability—affect real-world sensor responses. Therefore, understanding and defining biosensor performance metrics is essential to evaluate their accuracy, reliability, and suitability for analytical applications.

A biosensor's performance is characterized by its transducer, and in an ideal scenario with perfect materials, it would provide an absolute output value in response to a stimulus.

2.1 Calibration

The relationship between the physical measurement input signal (measurand) and the output signal is referred to as the transfer function. This transfer function can be used as a calibration curve. To calibrate a sensor, a series of standard physical inputs (solutions or gases) with known concentrations of the target analyte are applied. This process establishes a relationship between the output signal (y) produced by the sensor and the input (x): $y = f(x)$. The function, often referred to as a calibration function or curve, can take various forms, including simple linear relationships or nonlinear dependencies (e.g., logarithmic, exponential, or power functions). The calibration curve is constructed within the expected range of values to be measured, and it typically doesn't require extrapolations outside of this range to understand the biosensor's operational behavior.

A. Simonian et al., *Biosensors as Analytical Tools for the 21st Century*,
https://doi.org/10.1007/978-3-032-01726-0_2

2.1.1 *Construction of Calibration Curves Using Standard Solutions*

Standard solutions contain known concentrations of the analyte(s). Blank solutions, on the other hand, contain all the reagents and solvents used in the analysis but do not intentionally include the analyte.

Here's a step-by-step guide on constructing calibration curves using standard solutions:

1. **Prepare Known Samples**: Start by preparing known samples of the analyte, covering a range of concentrations that are expected for unknown samples. Measure the response of the analytical procedure for each of these standard solutions.
2. **Subtract Blank Values**: Subtract the average response of the blank samples from each measured standard's response to obtain a corrected value.
3. **Graph Data**: Create a graph that plots the corrected values against the concentration of the standard solutions. You can use the "method of least squares" procedure to find the best-fitting straight line through the linear portion of the data.
4. **Determine Unknown Concentrations**: To determine the concentration of an unknown sample, analyze the unknown along with a blank sample, subtract the blank's response to obtain the corrected value, and use this corrected value to determine the concentration based on your calibration curve.

Calibration is a crucial step in ensuring the accuracy and reliability of biosensor measurements.

2.2 Dynamic Range

The range within which a sensor can effectively operate is defined by the maximum and minimum values of the measurands, as illustrated in Fig. 2.1. This dynamic range of stimuli that a sensor can convert is commonly referred to as the span or full-scale input (FSI). While a sensor may still produce output signals outside of this range, doing so can result in erratic or inaccurate results. The FSI represents the highest possible input value that can be applied to the sensor without causing unacceptably large inaccuracies in the measurements. For sensors with broad and nonlinear response characteristics, the dynamic range of input stimuli is often expressed in a logarithmic measure. When the sensor's output does not precisely follow the transient response of the measured quantity, the error that occurs is referred to as dynamic error. *Dynamic error can arise when the sensor operates in conditions outside its ideal range or when it cannot respond quickly enough to rapid changes in the input stimuli* [1, 2].

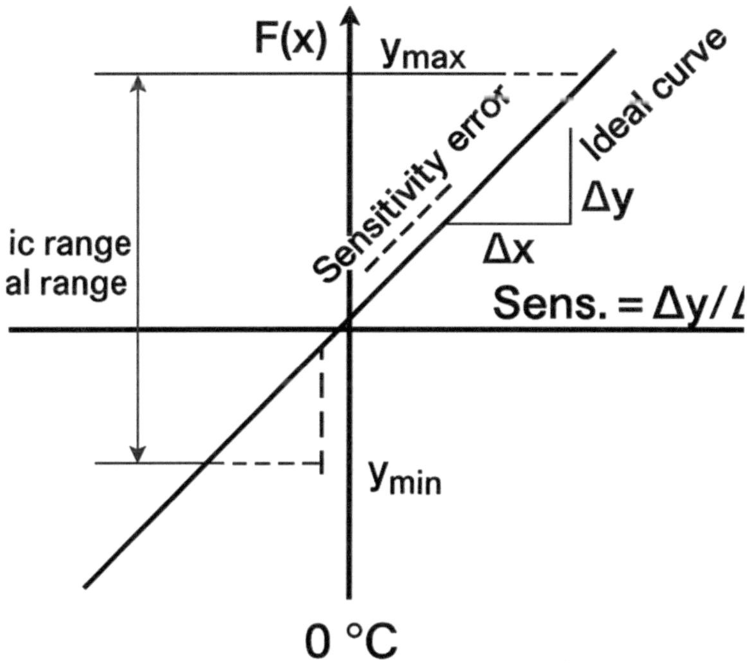

Fig. 2.1 Representation of the transfer/ideal curve, dynamic range, and sensitivity within a typical sensor response. The sensitivity error represents a deviation from the ideal slope of the sensor response

2.3 Sensitivity

The calibration curve, which closely approximates a straight line in a one-dimensional linear relationship, is expressed by the equation $y = mx + c$. Here, c represents the output signal at zero input signal, and m stands for the slope or sensitivity (S) across the entire range of x values to which it applies. The sensitivity, or scale factor, is essentially the gradient of the calibration curve, whether it remains constant or varies throughout the measurement range. Moreover, the sensor's sensitivity is also defined as the slope of the output characteristic curve ($\Delta y/\Delta x$ in the figure) or, more broadly, as the smallest change in the input physical parameter required to generate a detectable change in output. In certain sensors, sensitivity is defined as the alteration in input parameter necessary to produce a standardized output change. For electrochemical biosensors, electrodes are engineered with nanoparticles or carbon nanotubes that catalyze redox reactions. In these cases, sensitivity is determined by considering the electrode's surface area and can be calculated using the formula below:

$$\text{Sensitivity} = \frac{\text{Slope of calibration plot, } m\left(\mu\text{A mM}^{-1}\right)}{\text{Active Surface area, } A\left(\text{cm}^2\right)}.$$

2.4 Limit of Detection and Limit of Quantitation

"Sensitivity," "Limit of Detection," "Lower Limit of Detection," and "Limit of Quantitation" are terms used to characterize the minimum concentration of a measurand that can be accurately measured through an analytical procedure. In clinical applications, precise characterization of the analytical performance of every laboratory test is crucial to ensure accurate and dependable measurements of analytes, especially at very low concentrations. Similarly, biosensors undergo characterization and testing to determine their capabilities and limitations, ensuring they are suitable for their intended purposes. While "Sensitivity" is sometimes used interchangeably with "Limit of Detection," it's important to note that the Limit of Detection (LOD) may fall within a concentration range below the linear portion of an assay, where the calibration curve is no longer valid.

Despite their seeming similarity, these parameters have distinct definitions, and it's important not to confuse them.

Limit of Detection (LOD): This is the smallest concentration of analyte in a sample that can be reliably distinguished from zero, following the IUPAC definition. LOD can be calculated in three ways:

1. **Visual Evaluation**: Determined by visual inspection.
2. **Signal-to-Noise (S/N) Ratio**: Applicable to procedures that exhibit baseline noise. It involves comparing the signal from a low concentration of analyte to that of a blank.
3. **Based on Standard Deviation (S.D.) of Response and Slope**: Calculated using the standard deviation of the response and the slope of the calibration curve.

$$\text{Limit of Detection} = \frac{3.3\sigma\,(\text{S.D.})}{s(\text{slope})}.$$

In the calculation of the **Limit of Detection (LOD)**:

- s represents the slope of the calibration curve.
- σ stands for the standard deviation of the response, which can be obtained through various methods, including the standard deviation of blank response, the residual standard deviation of the regression line, or the standard deviation of the y-intercept of the regression line (Sy/x), which is essentially the standard error of estimate.

On the other hand, the **Limit of Quantitation (LOQ)** is defined as the lowest amount of analyte in a sample that can be quantitatively determined with appropriate precision and accuracy, following the definition by the International Council for Harmonization of Technical Requirements for Pharmaceuticals for Human Use (ICH). The calculation of LOQ is similar to that of LOD, but with slight variations in the formula [3–5].

$$\text{Limit of Quantitation} = \frac{10\sigma(\text{S.D.})}{s(\text{slope})}$$

2.5 Selectivity and Specificity

Selectivity is a term used to describe the ability of an analytical method or sensor to accurately and independently determine multiple analytes of interest simultaneously. It means that when there are several different substances present in a sample, the method or sensor can distinguish and measure each of them accurately without interference from the others.

Specificity, on the other hand, refers to the ability of an analytical method or sensor to accurately measure a single, individual component of a multicomponent system. In this case, the method or sensor is designed to focus exclusively on one particular analyte and provide an accurate measurement of that specific substance without being influenced or affected by the presence of other substances in the sample.

In summary, selectivity deals with the simultaneous measurement of multiple analytes, while specificity deals with the accurate measurement of a single analyte within a complex mixture. Both properties are important in analytical chemistry and sensor technology, depending on the specific application and analytical requirements.

2.6 Accuracy and Precision

Accuracy and precision are important concepts in measurement and sensor technology, and they are often used together to describe the quality of a sensor's performance:

1. **Precision**: Precision refers to the degree of reproducibility of measurements. In other words, it assesses how consistent and close to each other repeated measurements of the same quantity are. A highly precise sensor will produce very similar or nearly identical readings when the same measurement is taken multiple times. Precision is a measure of the sensor's ability to provide consistent results.

2. **Accuracy**: Accuracy, on the other hand, is a measure of how close the measured values are to the true or known values. It evaluates how well the sensor's measurements reflect the actual values of the quantity being measured. An accurate sensor provides measurements that are very close to the true values. Accuracy assesses the sensor's ability to provide correct results.

For example, a pressure of 150 mm Hg is applied to a sensor. Even if the applied pressure doesn't change, the output values from the sensor will vary considerably. Some subtle problems occur in the matter of precision when the true value and the sensor's mean value are not within a certain distance of each other (e.g., the $1-\sigma$ range of the statistician's normal distribution curve) [2, 6].

2.7 Repeatability

Repeatability or reproducibility is the ability of a sensor to represent the same value under identical conditions. This is similar to *accuracy*. However, certain sensors can be inaccurate and still give repeatable output signals. Reproducibility is expressed as a maximum difference between output readings as determined by two calibration cycles. It is usually represented as % of full-scale output.

$$\delta r = \frac{\Delta}{\text{FSO}} \times 100\%$$

Full-scale output (FSO) is the algebraic difference between the electrical output signals measured with maximum input stimulus and the lowest input stimulus applied. This must include all deviations from the ideal transfer function. Reproducibility can also be defined as a measure of the scatter or the drift in a series of observations or results performed over a period of time for the analyte concentrations within the usable range.

2.8 Linearity and Saturation

The *linearity* of the transducer is an expression of the extent to which the actual measured curve of a sensor departs from the ideal curve. Linearity is often specified in terms of *percentage of nonlinearity*, which is defined as:

$$\begin{aligned}\text{Nonlinearity (\%)} = {} & D\,\text{in (max) is the maximum input} \\ & \text{deviation/FSI is the maximum, full-scale input} \times 100\end{aligned}$$

The sensor operates in linearity under certain range of input stimulus, however upon further increase, its output signal no longer will be responsive. A future increase of stimulus does not produce a desirable output. It is said that the sensor exhibits a span-end nonlinearity or *saturation.*

2.9 Stability

The accuracy of the performance of the biosensor over a certain time period. The stability is measured upon consideration of the analyte concentration, the continuous or sequential contact of the biosensor with the analyte solution, temperature, pH, buffer composition, presence of organic solvents, and sample matrix composition [7, 8].

References

1. Joseph, J. C. *Sensors and Circuits: Sensors, Transducers, and Supporting Circuits for Electronic Instrumentation, Measurement, and Control.* (PTR Prentice Hall, 1993).
2. Pallas-Areny, R. & Webster, J. G. *Sensors and Signal Conditioning, Second Edition.* https://www.researchgate.net/publication/303748874 (2001).
3. Armbruster, D. A. & Pry, T. *Limit of Blank, Limit of Detection and Limit of Quantitation. Clin Biochem Rev* vol. 29 (2008).
4. Tholen, D. W. *Protocols for Determination of Limits of Detection and Limits of Quantitation: Approved Guideline.* (NCCLS, 2004).
5. *INTERNATIONAL CONFERENCE ON HARMONISATION OF TECHNICAL REQUIREMENTS FOR REGISTRATION OF PHARMACEUTICALS FOR HUMAN USE ICH HARMONISED TRIPARTITE GUIDELINE VALIDATION OF ANALYTICAL PROCEDURES: TEXT AND METHODOLOGY Q2(R1).*
6. Webster, J. G. *The Measurement, Instrumentation and Sensors Handbook.* (CRC Press, 1998). https://doi.org/10.1201/9781003040019.
7. Eggins, B. R. *Biosensors: An Introduction.* (Vieweg+Teubner Verlag, Weisbaden, 1996). https://doi.org/10.1007/978-3-663-05664-5.
8. Jonathan, M. C. and A. E. G. C. *Biosensors: A Practical Approach.* (Oxford; New York: Oxford University Press, 2004).

Chapter 3
Biorecognition Elements

3.1 Introduction

A central challenge in biosensor development is creating robust affinity-based methods that ensure both selectivity and sensitivity—the two essential pillars of analytical detection. Traditional biosensing platforms rely on highly specific biorecognition elements capable of detecting and identifying target analytes. High-affinity antibodies and nucleic acids have played foundational roles in immunoassays and immunosensors, particularly when integrated with electrochemical, optical, or mass-sensitive detection strategies. To meet diverse application needs, a variety of enhancements have been introduced, including labeling and label-free formats, nanomaterials, multifunctional matrices, and advanced structures such as carbon nanotubes.

However, despite the increasing affinity capabilities of conventional biosensors, the field faces challenges in meeting crucial demands such as heightened sensitivity, rapid detection, extreme selectivity, a robust biological foundation, calibration, in vivo measurements, regeneration capability, consistent performance, and ease of production. Nevertheless, recent advancements in this area have given rise to novel high-tech sensing systems, which exhibit remarkable potential. For example, aptamers, single-stranded DNA or RNA molecules that can specifically bind to target molecules with high affinity, making them valuable tools for biosensing applications. They offer distinct advantages over antibodies, including increased sensibility, stability, the ability for multi-detection, suitability for in vivo measurements, and the capacity for regeneration. Biosensors that employ aptamers as their biorecognition elements are commonly known as aptasensors.

The biomolecular recognition element is at the core of biosensing technology. Since the mid-1990s, advances in bioanalysis have significantly expanded our understanding of the development of detection and monitoring devices for various biologically sensitive elements. While a higher percentage of research documents have focused on detection instruments, the ideal sensing systems have truly emerged

A. Simonian et al., *Biosensors as Analytical Tools for the 21st Century*,
https://doi.org/10.1007/978-3-032-01726-0_3

through the development of biosensors. These biosensors incorporate a biorecognition element that is attached to the transducer, providing a sensitive read-out signal corresponding to the concentration of the analyte. Biorecognition elements exhibit unique specificity and sensitivity toward either a single analyte or a closely related group of analytes. Initially, these elements were isolated in laboratories from microorganisms, tissues, organelles, etc., but with the advent of emerging technology, they are now available as commercial products. Catalytic biosensors employ enzymes as biorecognition elements, leveraging their catalytic efficiency to detect a specific analyte (e.g., glucose biosensors specific to glucose) or a group of analytes (e.g., phenolic biosensors specific to various phenols). Receptors are also employed in biosensing due to their dual "receiving" and "sending" functions.

3.2 Naturally Inspired Biorecognition Elements

Naturally inspired biorecognition elements encompass natural targets, including transmembrane and soluble proteins, which exhibit the ability to bind to specific molecules known as ligands. This binding event initiates a specific cellular response and often involves conformational changes in the ligand-receptor interaction, triggering various metabolic events within cellular processes.

Initially, receptors were identified as promising biosensor elements for detecting a range of target ligands. However, over time, their practical use faced challenges due to their low yield, relative instability, labor-intensive isolation processes, and the cumbersome purification procedures required for membrane-associated proteins. These obstacles hindered their widespread adoption as sensing elements.

3.2.1 Enzymes: Protein Structure, Properties, and Kinetics

To grasp the nature of enzymes, it's essential to start by understanding proteins because enzymes are a subset of proteins. Proteins are intricate, large, and abundant organic molecules that play vital roles in every aspect of cell structure and function. They exhibit a wide range of properties, including structural and catalytic functions, owing to their diverse roles, despite their fundamental structural similarity.

Proteins are essentially biopolymers composed of amino acids. This designation stems from the fact that the amino group is bonded to the carbon atom adjacent to the carbonyl group in their structure. All proteins are composed of one or more lengthy chains of subunits known as amino acids, and it is the specific sequence of these amino acids that determines the physical and chemical characteristics of a protein. In nature, there are only 20 distinct amino acids, and each individual amino acid subunit is connected through amide linkages, commonly referred to as peptide bonds. Figure 3.1 provides an overview of the general structure of amino acid and protein.

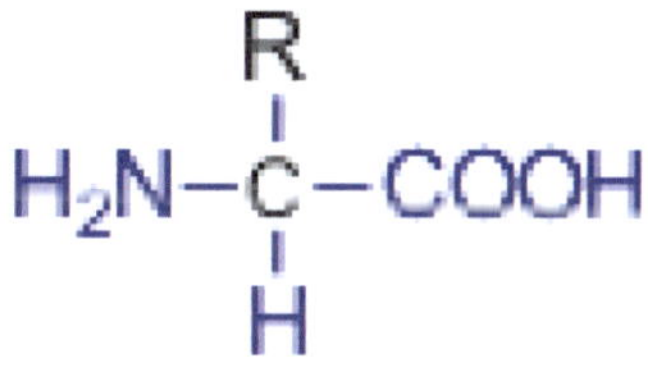

Fig. 3.1 General structure of an amino acid

The general structure of an amino acid consists of several key components:

1. **Amino Group (NH_2)**: Amino acids contain an amino group, which consists of two hydrogen atoms (H) bonded to a nitrogen atom (N). This group is often referred to as the "amino" part of the amino acid.
2. **Carboxyl Group (COOH)**: Amino acids also have a carboxyl group, which consists of a carbon atom (C) double-bonded to an oxygen atom (O) and single-bonded to a hydroxyl group (OH). This group is often called the "carboxyl" or "acid" part of the amino acid.
3. **Alpha Carbon (α-C)**: The central carbon atom in an amino acid is known as the alpha carbon (α-C). It is bonded to the amino group, the carboxyl group, a hydrogen atom, and an R-group (side chain). The R-group varies among different amino acids and gives each amino acid its unique properties.
4. **Hydrogen Atom (H)**: The alpha carbon of an amino acid is typically bonded to a single hydrogen atom.
5. **R-Group (Side Chain)**: The R-group, also known as the side chain, is a variable group of atoms that differs among the 20 naturally occurring amino acids. It confers specific chemical and physical properties to each amino acid and plays a crucial role in determining the protein's overall structure and function.

The variation in the R-group is what distinguishes one amino acid from another. Some R-groups are small and hydrophobic, while others are large and polar. The specific sequence of amino acids in a protein, dictated by the genetic code, determines the protein's unique three-dimensional structure and its biological function.

Amino acids are the building blocks of proteins, and they are indeed versatile molecules with various functions in living organisms. Here's a bit more information about amino acids:

1. **Peptide Bonds**: Peptide bonds are covalent bonds that link one amino acid to another. These bonds form between the amino group ($–NH_2$) of one amino acid and the carboxyl group (–COOH) of another amino acid, releasing a molecule of water in the process. The resulting linkage is called a peptide bond, and a chain of amino acids linked by peptide bonds is known as a polypeptide chain. When the polypeptide chain folds into a specific three-dimensional structure, it becomes a functional protein.
2. **Variety of Amino Acids**: While there are 20 commonly occurring amino acids in proteins, more than 150 amino acids can be found in nature. These amino acids may have different chemical properties and functions in the body.

3. **Alpha-Amino Acids**: The amino acids found in proteins are often referred to as alpha-amino acids because they have both an amino group ($-NH_2$) and a carboxylic acid group (–COOH) as substituents on the same carbon atom, known as the alpha carbon (α-C). This alpha carbon is also bonded to a hydrogen atom and the R-group, which varies among different amino acids.
4. **Proline Exception**: Proline is unique among the 20 common amino acids because it has an imino group (–NH) in its side chain, making it the only amino acid with two nitrogen atoms in its side chain. This structural difference affects its role in protein structure and function.
5. **Diverse R-Groups**: The R-group, or side chain, of each amino acid is what sets it apart from the others. These side chains can be simple and nonpolar, polar and hydrophilic, or charged and hydrophilic/hydrophobic. The specific chemical properties of the R-groups influence how amino acids interact with each other and their surroundings, ultimately determining the protein's function.

The sequence of amino acids in a protein is determined by the genetic code and plays a critical role in the protein's structure and function. Even a slight change in the sequence of amino acids can lead to significant alterations in protein structure and function, which can have profound effects on an organism's health and physiology.

General Properties

Amino acids indeed have unique chemical properties due to their structure, including amino ($-NH_2$) and carboxylic acid (–COOH) groups that can readily ionize. Here's a more detailed explanation of amino acid ionization and the concept of the isoelectric point (pI):

1. **Amino Acid Structure**: As already mentioned, amino acids consist of several functional groups bonded to a central carbon atom (Cα). These functional groups include a hydrogen atom (H), an amino group ($-NH_2$), a carboxyl group (–COOH), and the R-group, which varies among different amino acids.
2. **pK Values**: Amino acids have specific pK (acid dissociation constant) values for the ionization of their functional groups. The pK values are a measure of the acidity or basicity of a compound. Amino acids have two key pK values:

   - The pK value for the carboxyl group (–COOH) is typically around 2.2. When the carboxyl group loses a proton (H^+), it becomes deprotonated and forms the carboxylate ion (–COO–).
   - The pK value for the amino group ($-NH_2$) is generally around 9.4. When the amino group gains a proton, it becomes protonated and forms the ammonium ion ($-NH_3^+$).

3. **Isoelectric Point (pI)**: The isoelectric point (pI) of an amino acid is the pH at which the molecule carries no net electrical charge. At this point, the amino group is protonated ($-NH_3^+$), and the carboxyl group is deprotonated (–COO–), resulting in a zwitterion—a molecule with both positive and negative charges. The pI value is unique for each amino acid and can be calculated based on its pK values.

4. **Titration Curve**: The titration curve for an amino acid illustrates how its charge changes with pH. It shows the transition from a positively charged form (low pH) to a negatively charged form (high pH) with the isoelectric point (pI) at the pH where the net charge is zero. The shape of the titration curve depends on the specific pK values of the amino acid.

Understanding the ionization behavior of amino acids is crucial in biochemistry and biology, as it plays a fundamental role in protein structure, function, and interactions. Proteins, which are composed of amino acids, can have complex three-dimensional structures and charge distributions that depend on the pH of their environment. This pH-dependent behavior is essential for processes such as enzyme activity, protein folding, and molecular recognition in biological systems.

The titration curve of alanine (Fig. 3.2) illustrates how the charge on the alanine molecule changes as the pH of the solution varies. It's a fundamental concept in understanding the behavior of amino acids and their isoelectric point (pI). The Henderson–Hasselbalch equation is indeed useful for calculating the pH at different stages of protonation and deprotonation.

For neutral species with two ionizable groups, the H–H equation can be expressed as:

$$\mathrm{pI} = \frac{1}{2}\left(\mathrm{pK}_i + \mathrm{pK}_j\right).$$

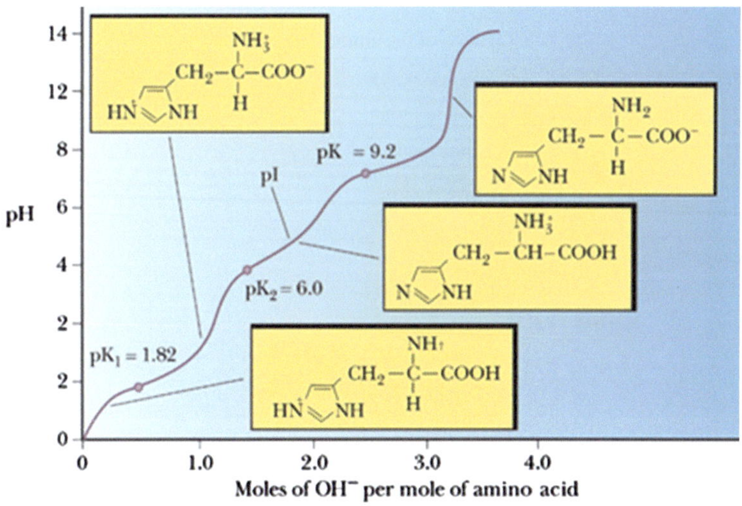

Fig. 3.2 Titration curve of histidine illustrating the relationship between pH and the addition of hydroxide ions (OH^-) per mole of amino acid. The curve displays three distinct buffering regions corresponding to the ionizable groups in histidine: the carboxyl group ($pK_1 = 1.82$), the imidazole side chain ($pK_2 = 6.0$), and the amino group ($pK = 9.2$). The isoelectric point (pI) is marked where the molecule exists predominantly in its zwitterionic form. Molecular structures at different pH values indicate the protonation states of histidine throughout the titration [1]

To summarize:

- At low pH (acidic conditions), both the amino group (–NH_2) and the carboxyl group (–COOH) are fully protonated, resulting in a net positive charge on alanine.
- As pH increases, the amino group begins to deprotonate, and there's a transition to the zwitterionic form where the positive charge on the amino group starts to balance the negative charge on the carboxyl group.
- At the isoelectric point (pI), which is around pH 2.35 for alanine, there is no net electric charge, and alanine exists as a zwitterion.
- As pH continues to increase beyond the pI, the carboxyl group starts to deprotonate, leading to a net negative charge on alanine.

This behavior is consistent with the behavior of amino acids in solution and is crucial in understanding the properties of proteins and enzymes, as the overall charge of these biomolecules can influence their structure and function. The Henderson–Hasselbalch equation helps calculate the pH at various stages of protonation, making it a valuable tool in biochemistry and analytical chemistry.

In Table 3.1, amino acids are categorized based on their side chains, also known as R-groups. There are several distinct groups of amino acids:

1. Aliphatic Amino Acids: These amino acids have aliphatic side chains, which can include two-carbon side chains and branched chains.

Table 3.1 Classification of amino acids

Class	Name of the amino acids	R-group
Aliphatic	Glycine (Gly)	–H
	Alanine (Ala)	–CH_3
	Valine (Val)	–$CH_2CH(CH_3)_2$
	Leucine (Leu)	–$CH_2CH(CH_3)_2$
	Isoleucine (Ilu)	–$CH_2CH(CH_3)_2$
Hydroxyl or sulfur-containing	Serine (Ser)	–CH_2OH
	Cysteine (Cys)	–CH_2SH
	Threonine (Thr)	–$CH_2CH(OH)CH_3$
	Methionine (Met)	–CH_2CH_2–S–CH_3
Cyclic	Proline (Pro)	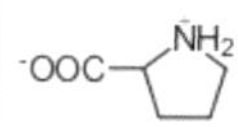
Aromatic	Phenylalanine (Phe), tyrosine (Tyr), tryptophan (Trp)	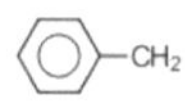 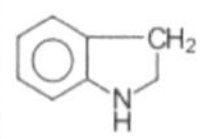

2. Amino Acids Classified by OH Groups: Certain amino acids, like threonine and serine, are categorized based on the presence of OH groups in their side chains.
3. Cysteine: Cysteine is a unique amino acid characterized by its sulfhydryl group (–SH). It forms covalent disulfide bridges (–S–S–) with other protein subunits containing –SH groups and can also bind to thiolated solid supports.
4. Aspartic Acid and Glutamic Acid: These amino acids have COOH side groups and are considered acidic amino acids. They can form amide conjugates, resulting in aspartate and glutamate.
5. Basic Amino Acids: Lysine, histidine, and arginine have NH_2 as their R-group. These are classified as basic amino acids. The overall charge of proteins can vary depending on the presence of acidic or basic groups in the constituent amino acids. Changes in pH within the protein's microenvironment can alter its charge.
6. Chemical Properties: Amino acids exhibit distinct chemical properties due to the type and arrangement of atoms in their side chains. Functional groups such as –OH, $-NH_2$, –COOH, and –SH contribute to these properties and enable interactions with external molecules, atoms, ions, or side chains of other proteins. These interactions can include hydrogen bonds.
7. Polarity: Amino acids with acidic or basic functional groups can exhibit polarity and act as either acids or bases within pH ranges characteristic of their surrounding environment.
8. Nonpolar Amino Acids: Amino acids like glycine, valine, leucine, isoleucine, phenylalanine, tryptophan, and proline lack functional groups and are nonpolar. Despite this, they can readily interact with other molecules, often forming hydrogen bonds.

The diverse chemical properties of amino acids play a crucial role in the structure and function of proteins and their interactions with the surrounding biological environment.

Most amino acids possess four distinct chemical groups attached to a central carbon atom, except for glycine. Consequently, they exhibit an asymmetric structure and exist in both D and L stereoisomeric forms. It's worth noting that amino acids found in ribosomes are exclusively in the L-isomeric form, while D-isomers are prevalent in bacterial cells.

Protein Structure

Amino acids can be connected in various combinations to form chains consisting of two or more units. When two amino acids are linked together, the resulting structure is known as a "dipeptide," while chains containing three or more units are referred to as "polypeptides." The linkage between two amino acids is established through a covalent bond known as the "peptide bond" or "peptide linkage." This bond is formed through a condensation reaction between the NH_2 group of the first amino acid and the COOH group of the second amino acid, as illustrated in Fig. 3.3.

As additional amino acids are added to the free COOH end of the dipeptide through peptide linkages, the dipeptide continues to grow, forming longer chains.

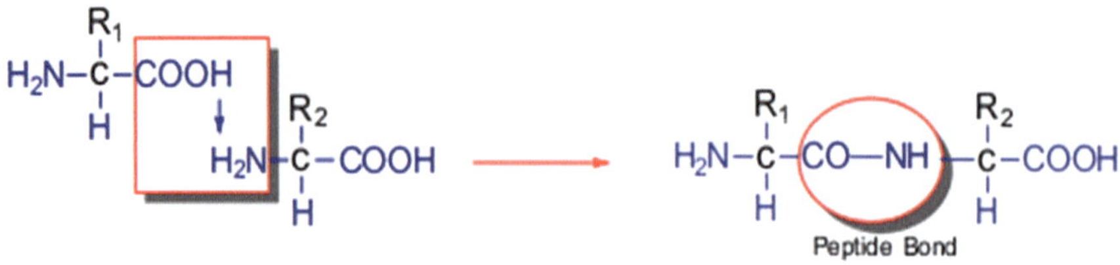

Fig. 3.3 Formation of a peptide bond between two amino acids. The carboxyl group (–COOH) of one amino acid reacts with the amino group (–NH_2) of another, resulting in the release of a water molecule and the formation of a covalent bond (peptide bond) between the carbonyl carbon and the nitrogen atom. The newly formed peptide bond is highlighted in red [2]

Consequently, the linear sequence of amino acids in a protein is referred to as its primary structure.

The secondary structure of a protein pertains to the spatial arrangement of the atoms in the polypeptide backbone, disregarding the specific orientation of the side chains.

In the late 1930s, Linus Pauling and Robert Corey analyzed the peptide structure. The spatial arrangement of atoms in the peptide bond involves a twisting of the bond, and the resistance to this twisting plays a crucial role in determining the possible folding conformations of polypeptide chains. Consequently, by examining the rotational freedom of bonds around certain degrees on either side of the carbon atoms, two distinct ordered arrangements can be observed for an amino acid sequence: α-helices and β-sheets, as illustrated in Fig. 3.4a, b.

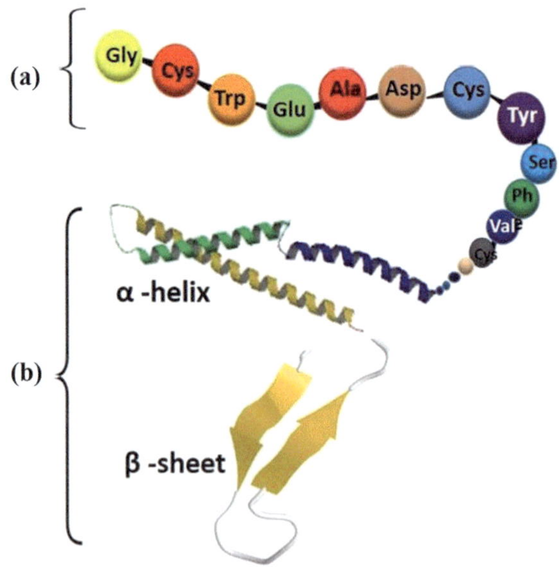

Fig. 3.4 Illustration of protein structure. **a** Linear sequence of amino acids representing the primary structure, with each colored sphere labeled by its respective amino acid abbreviation. **b** Secondary structure of the protein, showing the formation of α-helices and β-sheets from the linear chain. These structural motifs result from hydrogen bonding patterns between backbone atoms in the polypeptide chain [3]

There are three primary low-energy backbone arrangements that can be found: the right-hand helix (α-helix), antiparallel β-sheets, and the less common left-hand helix. The stability of these secondary structures is largely maintained by hydrogen bonds formed between the backbone atoms, except in the case of glycine, which lacks a side chain.

Tertiary structures encompass the three-dimensional arrangements of an entire polypeptide chain. The specific distribution of α-helices and β-sheets within a protein is determined by its amino acid sequence, under specific conditions of pH and temperature.

The formation and determination of tertiary structure involve several factors, including:

- The presence and location of disulfide linkages.
- The presence of proline residues.
- The positioning of hydrogen bonds.
- Interactions between positively and negatively charged side groups.
- van der Waals forces.
- Hydrophobic interactions, both among polar and nonpolar residues.

These elements collectively contribute to the overall three-dimensional conformation of a protein, known as its tertiary structure.

The van der Waals forces play a substantial role, accounting for 75% of the minimum energy required for globular protein folding. This contribution imparts significant rigidity to protein structures, rendering them akin to most organic solids. Protein structures involve intricate 3D folding patterns held together by relatively small amounts of energy acting on the amino acid domains.

Under varying conditions, such as changes in temperature, pH, and concentrations, proteins can undergo structural perturbations due to energy input from the surroundings. This may lead to the unfolding and subsequent refolding of proteins into random shapes.

Complex proteins often exhibit a quaternary structure. Hemoglobin and antibody molecules serve as notable examples, consisting of multiple polypeptide chains. These molecules are composed of four polypeptide chains, and the specific arrangement of these chains, intermingling in a particular pattern, defines their quaternary structure. Figure 3.5 provides a comprehensive overview of various protein structures, highlighting the interactions responsible for protein folding.

Protein Complex Structures

Proteins exhibit a wide array of reactive groups within their side chains, allowing them to engage in diverse biomolecular and biochemical interactions with other complex molecules, such as carbohydrates or lipids. When proteins covalently bind with carbohydrate units, they form glycoproteins. Glycoproteins play crucial roles as enzymes, antibodies, recognition and receptor molecules on cell surfaces, and as components of extracellular structures like collagen. Similarly, when proteins bind covalently with lipid units, they give rise to lipoproteins, which are found extensively in both eukaryotic and prokaryotic cells.

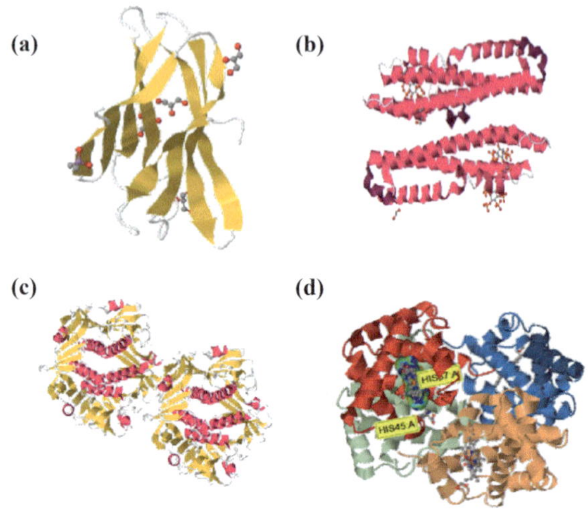

Fig. 3.5 **a** Depicts a 3D model of β-sheets, representing a secondary structure of a protein. **b** Shows α-helix structures, another secondary structure of a protein. **c** Represents the tertiary structure of a protein. **d** Illustrates the quaternary structure of hemoglobin, featuring four subunits (2 α-helices and 2 β-sheets) in a space-filling model

Enzymes

Enzymes are globular protein molecules that serve as catalysts for specific biochemical reactions. These catalysts lower the activation energy required for reactions, thereby accelerating the reactions without altering the final products or the equilibrium constants of the reactions. The catalytic function of enzymes involves the conversion of defined substrate molecules into different molecules, referred to as products.

Specifically, the substrate interacts with the enzyme through a reversible binding process, forming an enzyme-substrate complex. This complex is subsequently converted irreversibly into an enzyme-product complex, leading to the release of the product through a reversible unbinding process. This process is elaborated in detail in Fig. 3.6.

Enzymes are systematically categorized based on their functions and catalytic reactions using the Enzyme Commission (EC) number classification. There are seven main classes of enzymes:

1. EC 1—Oxidoreductases: These enzymes facilitate redox reactions, involving the transfer of electrons between molecules, leading to oxidation and reduction processes.

$$\begin{array}{c} \text{Enzyme} + \text{Substrate} \leftrightharpoons \text{Enzyme Substrate Complex} \\ \downarrow \\ \text{Enzyme} + \text{Product} \leftrightharpoons \text{Enzyme Product Complex} \end{array}$$

Fig. 3.6 Schematic description of enzymatic conversion of substrate into product through the enzyme-substrate complex

2. EC 2—Transferases: Transferases are responsible for transferring specific functional groups, such as methyl or phosphate groups, from one molecule to another.
3. EC 3—Hydrolases: Hydrolases catalyze hydrolysis reactions, breaking various chemical bonds by the addition of water molecules.
4. EC 4—Lyases: Lyases cleave various chemical bonds by mechanisms other than hydrolysis and oxidation. They often result in the formation of new double bonds or the addition of groups to molecules.
5. EC 5—Isomerases: Isomerases catalyze isomerization reactions, leading to changes in the spatial arrangement of atoms within a single molecule. This class is involved in structural isomer transformations.
6. EC 6—Ligases: Ligases join two molecules together by forming covalent bonds. This class of enzymes is often responsible for synthesizing larger molecules from smaller ones.
7. EC 7—Translocases[1]: Translocases are involved in catalyzing the movement of ions or molecules across biological membranes. They play a crucial role in membrane transport processes.

The Enzyme Commission number system provides a standardized way to classify and categorize enzymes based on their specific roles and biochemical activities in various biological processes.

Enzymes and Active Sites: An active site is the region on an enzyme where a substrate binds, forming a pocket or groove. This interaction initiates a chemical reaction. The active site's structure is determined by the enzyme's tertiary and quaternary levels of protein organization. While it is not rigid, it is highly specific, recognizing only one substrate. When the active site binds to substrate molecules, it induces a slight change in the enzyme's shape to facilitate the reaction, as depicted in Fig. 3.7. To understand enzyme-substrate binding, two models have been proposed.

1. **Lock and Key Model**: In this model, the enzyme and substrate have specific complementary shapes that fit precisely. The substrate (key) fits into the active site (keyhole) of the enzyme (lock), like a key fitting into a lock. It emphasizes the strict specificity of the interaction.
2. **Induced Fit Model**: This model describes how the active site of the enzyme can adjust its shape slightly to accommodate the substrate. It continues to change until the substrate is fully bound, determining the final shape and charge. Unlike the Lock and Key Model, the Induced Fit Model highlights the flexibility of enzyme structures.

Enzyme Cofactors

1. Enzyme cofactors, such as coenzymes and metal ions, are vital for regulating protein activity. Often, enzymes require these cofactors to function effectively. Examples of cofactors include essential metals like calcium, magnesium, and zinc, which are necessary for enzymes like carboxypeptidase.

[1] It is still under final approval by Enzyme Commission process.

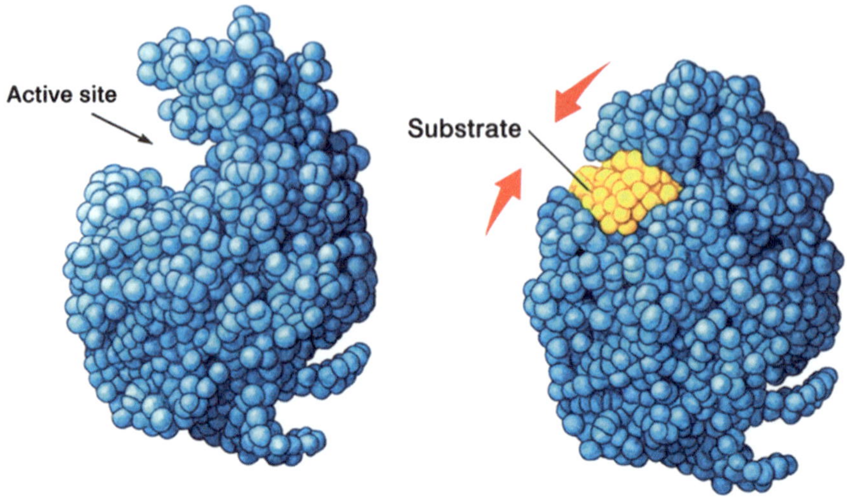

Fig. 3.7 Molecular model of an enzyme showing the active site and substrate binding. The left panel highlights the enzyme's active site (indicated by an arrow), a specific region where catalysis occurs. The right panel shows the substrate (in yellow) bound within the active site, illustrating the enzyme-substrate complex. This interaction is fundamental to enzymatic specificity and catalytic activity [4]

2. One important organic cofactor is nicotinamide adenine dinucleotide phosphate (NADP), frequently used in enzyme-based sensor systems. NADP plays a key role in electron transfer reactions, as seen in glucose-6-phosphate dehydrogenase (Fig. 3.8).
3. Similarly, nicotinamide adenine dinucleotide (NAD) is involved in numerous enzymatic reactions and exists in two forms: NAD+ and NADH. NADH acts as a reducing agent by donating electrons. It transforms from NAD+ after accepting electrons from other molecules. These changes can be monitored, often by observing alterations in optical properties or oxidation–reduction states, such as NAD/NADH.
4. Cofactors are pivotal in many enzymatically catalyzed reactions and undergo chemical transformations during the reaction process.
5. This revised version maintains the essential information while presenting it in a more structured and concise manner.

Enzyme Activity and Activation Energy. What Is Enzyme Activity and Specific Activity?

Enzyme Activity: Enzyme activity is the rate at which an enzyme converts its substrates into products within a specific timeframe. It is typically measured in "Units," where one unit represents the amount of enzyme that converts 1 μmol of substrate into product in 1 min. This measurement is taken under conditions where the substrate is not limited, and the enzyme is purified.

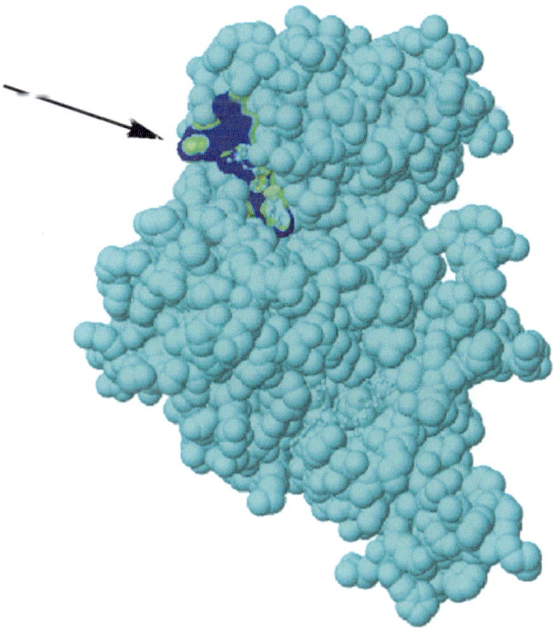

Fig. 3.8 X-ray structure of an enzyme complexed with NADP cofactor. This space-filling model depicts the interaction of glucose-6-phosphate dehydrogenase with its cofactor

$$1\ \text{Enzyme Unit}\ (U) = 1\ \mu\text{mol}/\text{min}$$

One Unit of Glucose Dehydrogenase Enzyme: One unit of glucose dehydrogenase enzyme catalyzes the oxidation of 1.0 μmol of β-D-glucose to D-glucono-lactone or D-galactose to D-galactonate per minute. This reaction occurs at pH 7.0 and 37 °C in the presence of the cofactor NADP+.

Specific Activity: Specific activity is the enzyme's activity per milligram of total volume (mL) of the enzyme. When enzymes are extracted from crude tissue cells and body fluids, specific activity serves as a measure of enzyme purity within the mixture. Specific activity (denoted as, or simply referred to as specific activity) can be calculated as follows:

$$\text{Unit}/\text{mg protein} = \frac{\text{units}/\text{mL (enzyme)}}{\text{g (protein)}/\text{mL (enzyme)}}.$$

Effect of Temperature on Enzyme Denaturation: Enzymes maintain their native structure through various interactions, including hydrogen bonds, van der Waals forces, electrostatic interactions, and chemical cross-links like disulfide bonds. However, these weak non-bonding forces can be disrupted by temperature changes. Heating of enzymes leads to a rapid alteration in their conformational properties over a narrow temperature range. This can increase enzyme activity by promoting collisions between active sites. Beyond a certain temperature threshold, the entire polypeptide chain unfolds and denatures cooperatively. For human enzymes, the optimal temperature typically falls in the range of 35–40 °C, beyond which enzyme activity decreases.

Effect of pH: Changes in pH alter the ionization states of amino acid side chains, affecting the distribution of charges within a protein. Each enzyme has its optimal pH, where its activity is maximized for catalytic reactions. Deviating from this optimal pH leads to a decrease in enzyme activity.

Effect of Substrate Concentration: As substrate concentration increases, the reaction rate also increases, assuming a fixed enzyme concentration. However, there is a point of saturation, where further increases in substrate concentration do not affect the reaction rate. At this point, all available enzyme active sites are occupied, and the substrate is efficiently converted to products. Excess substrate can't bind to additional active sites, which results in no further increase in the reaction rate.

Enzyme Kinetics

Louis Michaelis and Maud Menten's theory postulates the existence of an enzyme-substrate (ES) complex, with this complex being in rapid equilibrium with free enzyme. According to their theory, the breakdown of ES to yield products (*P*) is assumed to be slower than two other key processes.

The formation of ES, represented by the reaction: $E + S \rightleftharpoons \mathrm{ES}$, with rate constants k_1 (forward) and k_{-1} (backward).

The breakdown of ES to regenerate free enzyme (*E*) and substrate (*S*), represented by the reaction: $\mathrm{ES} \rightarrow P + E$, with a rate constant k_2.

$$E + Sk_1 \rightleftharpoons k_{-1}\mathrm{ES}k_2 \rightarrow P + E. \tag{3.1}$$

Here, *E*, *S*, ES, and *P* represent the enzyme, substrate, enzyme-substrate complex, and products, respectively. When the substrate concentration increases to a level where it saturates the enzyme to form ES, the second reaction is regarded as the rate-limiting step, causing the overall rate to become insensitive to further increases in substrate concentration. The rate constants k_1 and k_{-1} correspond to the forward and backward reactions of ES complex formation, respectively, while k_2 represents the rate constant for the forward reaction of the rate-determining step.

The rate of formation can be expressed as the product of the rate constant for the rate-limiting step and the concentration of the enzyme-substrate complex. Therefore, the velocity (v) or rate can be given as:

$$v = k_2[\mathrm{ES}]. \tag{3.2}$$

The rate of ES formation is represented as $[\mathrm{ES}] = k_1 * [E][S]$, and similarly, the rate of breakdown is denoted as $[\mathrm{ES}] = k_{-1} * [\mathrm{ES}]$.

Under physiologically common conditions, the substrate is present in excess compared to the enzyme. We can observe that ES formation remains constant until the substrate is consumed. Therefore, the rate of ES formation is equal to the rate of

ES breakdown, which then leads to:

$$k_1([E_{\text{total}}] - [\text{ES}])[S] = k_{-1}[\text{ES}] + k_2[\text{ES}]. \tag{3.3}$$

Further rearranging gives:

$$\begin{aligned} k_1([E_{\text{total}}][S] - k_1[\text{ES}])[S] &= k_{-1} + k_2[\text{ES}]k_1[E_{\text{total}}][S] \\ &= (k_1[S] + k_{-1} + k_2)[\text{ES}]. \end{aligned}$$

Letting $[E] = [E_{\text{total}}] - [\text{ES}]$ and rearranging yields:

$$\frac{[E_{\text{total}}] - [\text{ES}][S]}{[\text{ES}]} = \frac{k_2 + k_{-1}}{k_1}.$$

The Michaelis constant, K_{M} is defined as:

$$K_{\text{M}} = \frac{k_2 + k_{-1}}{k_1}.$$

Inserting K_{M} into the equation, it can be rearranged to give:

$$K_{\text{M}}[\text{ES}] = (E_{\text{total}} - [\text{ES}][S])$$

$$[\text{ES}] = \frac{[E_{\text{total}}][S]}{K_{\text{M}} + [S]}.$$

The expression for initial velocity is:

$$v = k_2[\text{ES}] = \frac{k_2[E_{\text{total}}][S]}{K_{\text{M}} + [S]}.$$

The maximum velocity of a reaction, often denoted as V_{max}, is achieved when the enzyme is saturated with its substrate. At this point, every active site of the enzyme is occupied by the substrate. Therefore, adding more substrate won't increase the rate of reaction any further. This is a key concept in enzyme kinetics and is often used to characterize the efficiency and capacity of enzymes. The Michaelis–Menten equation, which describes the rate of enzymatic reactions, is based on this principle.

$$V_{\text{max}} = k_2[E_{\text{total}}].$$

Combining these equations, we obtain the expression for the Michaelis–Menten equation, and K_{M} is denoted as the Michaelis–Menten constant.

$$v = \frac{V_{\text{max}}[S]}{K_{\text{M}} + [S]}.$$

Understanding Terms

$\boldsymbol{K_M}$: K_M is a constant derived from rate constants. K_M is under true Michaelis–Menten conditions, an estimate of the dissociation constant of E from S. A low K_M refers to strong binding (i.e., only a small amount of substrate is needed to saturate the enzyme); high K_M refers to weak binding (i.e., a lot of substrate should be present to saturate the enzyme).

$\boldsymbol{V_{max}}$: *The theoretical maximal velocity.*

$\boldsymbol{V_{max}}$ is indeed a constant representing the theoretical maximal rate of the reaction, but it is never achieved. Achieving V_{max} would require that all enzyme molecules are tightly bound with substrate, which is not practically feasible. V_{max} is asymptotically approached as substrate concentration is increased.

The turnover number ($\boldsymbol{k_{cat}}$), also referred to as the molecular activity of the enzyme, measures the enzyme's maximal catalytic activity. It quantifies the number of substrate molecules converted to product per enzyme molecule per unit of time when the enzyme (E) is saturated with substrate. This parameter provides insight into the efficiency of an enzyme in catalyzing a specific reaction.

If the M–M model fits, $k_2 = k_{cat}$, then $k_{cat} = \frac{V_{max}}{E_{total}}$.

The values of k_{cat} can indeed vary significantly between different enzymes and can range from less than 1 per second to many millions per second. For instance, if we compare the turnover numbers for catalase and lysozyme, we find that catalase has a very high turnover number of 40,000,000 per second, indicating it is highly efficient at catalyzing its specific reaction. In contrast, lysozyme has a much lower turnover number of 0.5 per second, indicating it catalyzes its specific reaction at a slower rate. These values highlight the diverse catalytic efficiencies of enzymes in different biological processes.

Catalytic Efficiency of an Enzyme—Catalytic Efficiency

$$\text{Catalytic Efficiency} = \frac{k_{cat}}{K_M}$$

Catalytic efficiency is indeed an estimate of "how perfect" an enzyme is in catalyzing a specific reaction. It represents an apparent second-order rate constant, which indicates how efficiently the enzyme and substrate interact when the substrate concentration (S) is low [4].

It's important to note that catalytic efficiency is constrained by the diffusion limit. The diffusion limit represents the maximum rate at which enzyme (E) and substrate (S) can collide and interact. Enzymes cannot catalyze reactions faster than this limit because it depends on the physical rate of diffusion. Enzymes that approach the diffusion limit are considered highly efficient in converting substrate to product, while those with lower catalytic efficiencies operate at a slower pace relative to the diffusion limit.

3.2.2 *Molecular Diagnostics: Nucleic Acid Structure, Kinetic and Affinity Principles*

Molecular diagnostics refers to the collection of analytical methods, including biosensors, used to analyze biological markers in both the genome and proteome. This field examines how cells express their genes as proteins by applying principles of molecular biology to medical testing. These techniques can diagnose and monitor diseases, assess risk, and determine the most effective therapies for individual patients. Many of these analytical methods rely on affinity interactions between nucleic acids or antibodies and their respective analytes.

Nucleic Acids: Components and Structure

Nucleotides are composed of a five-carbon sugar, known as a pentose, covalently linked to a nitrogenous base at the C1 position. These nitrogenous bases are planar, aromatic, heterocyclic molecules, primarily derivatives of purines or pyrimidines. In the case of DNA and RNA, their chemical structures consist of linear polymers of nucleotides, with phosphate groups forming bridges between the 3′ and 5′ positions of successive sugar residues (refer to Fig. 3.9).

Double Helical DNA

The two strands of DNA are connected through base pairing, with each base having a specific partner: adenine (A) pairs with thymine (T), and guanine (G) pairs with cytosine (C). This DNA base pairing, or complementarity, is illustrated in Fig. 3.10.

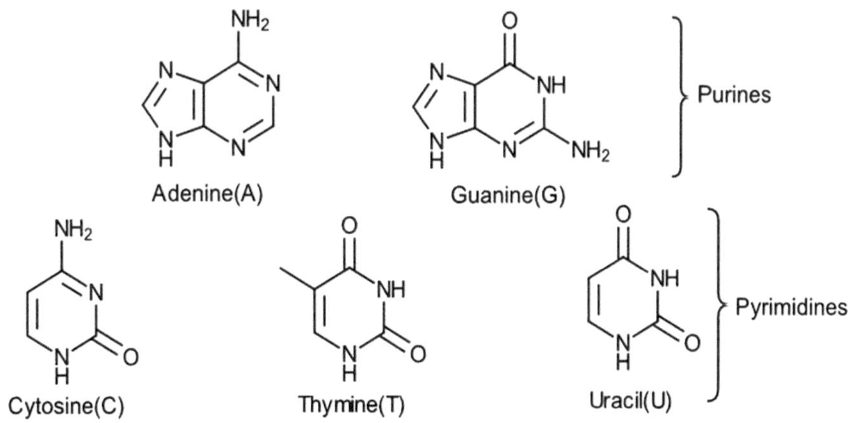

Fig. 3.9 Chemical structures of the five major nitrogenous bases found in nucleic acids. Adenine (A) and guanine (G) are purines, characterized by a double-ring structure. Cytosine (C), thymine (T), and uracil (U) are pyrimidines, each with a single-ring structure. Thymine is present in DNA, while uracil is found in RNA, replacing thymine. These bases are essential components of nucleotide sequences that encode genetic information [5]

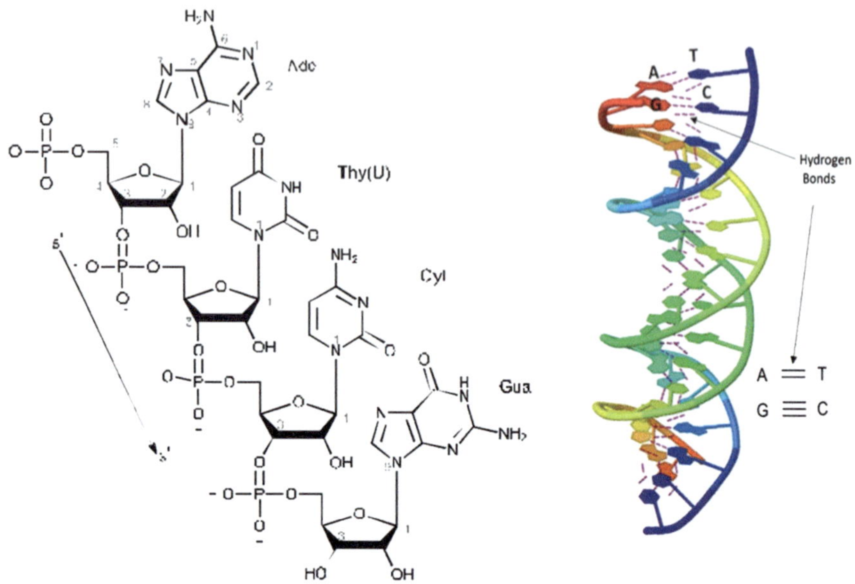

Fig. 3.10 Molecular representations of DNA structure. (Left) Chemical structure of deoxyribonucleic acid (DNA) showing the nucleotide backbone, sugar-phosphate groups, and nitrogenous bases. (Right) 3D double helix model of DNA, highlighting base pairing between adenine (A)–thymine (T) and guanine (G)–cytosine (C) via hydrogen bonds, which stabilize the helical structure. The image illustrates the complementary and antiparallel nature of DNA strands [5]

The DNA strands are read in a specific direction, from the top, referred to as the 5′ or "five prime" end, to the bottom, known as the 3′ or "three prime" end. These two strands intertwine in a right-handed helix. The backbones of the strands run antiparallel to each other, with one going from 5′ to 3′ and the other going from 3′ to 5′. They have opposite polarities.

The sugar-phosphate backbone is positioned on the outside of the helix, while the bases are paired and stacked on the inside. This geometric arrangement strongly favors A-T and G-C base pairs, ensuring the faithful duplication of the DNA molecule.

5′ T–T–G–A–C–T–A–T–C–C–A–G–A–T–C 3′
3′ A–A–C–T–G–A–T–A–G–G–T–C–T–A–G 5′

Denaturation and Renaturation: When a solution of duplex DNA is heated above a specific temperature, the native structure of the DNA collapses, and its two complementary strands separate. They adopt a flexible and rapidly fluctuating conformational state known as a random coil. Denaturation can also be induced by factors such as pH and high salt concentrations. The temperature at which half of the hydrogen bonds are broken (half-denatured) is referred to as the Tm or characteristic melting temperature.

Central Dogma of Life: In 1958, Francis Crick succinctly summarized the fundamental processes in molecular biology with a concept he termed the central dogma. According to this framework, DNA directs its own replication and transcription to produce RNA. In turn, RNA guides translation to synthesize proteins. In this context, "transcription" signifies the transfer of genetic information from DNA to RNA, with the language encoding the information remaining consistent based on base sequences. Conversely, "translation" denotes the transfer of information from RNA to proteins, with the language shifting from base sequences to amino acid sequences.

DNA Replication: Every time a cell divides into two daughter cells, the entire DNA molecule must be duplicated, a process known as replication. During replication, the DNA double helix unwinds, and its two strands separate. A replication bubble forms and progresses along the DNA molecule in both directions. Each DNA strand serves as a template for the synthesis of a new complementary strand. The primary responsibility for replication lies with a group of enzymes known as DNA polymerases. These enzymes recruit free nucleotides and pair them by base pairing with the complementary nucleotides of the parent strand. Consequently, during each cell division, DNA polymerase duplicates the entire DNA molecule, ensuring that one copy is passed to each daughter cell.

Restriction Endonucleases: Specific palindromic sequences scattered throughout a DNA molecule serve as recognition sites for restriction endonucleases. These enzymes play a pivotal role in molecular biology by cleaving the sugar-phosphate backbone of DNA at these recognition sites. The cleavage results in either a staggered cut or an unpaired single-strand break. Importantly, DNA molecules or restriction fragments cleaved by the same restriction enzyme possess complementary single-strand ends. These fragments can be covalently joined together under renaturation conditions, facilitated by DNA ligase.

Understanding these processes is fundamental to numerous molecular biology techniques, including DNA cloning, gene mapping, and the study of gene function.

PCR Amplification

PCR, short for the polymerase chain reaction, is an iterative method that employs a pair of oligonucleotide primers, deoxynucleotide-5′-triphosphates, and a heat-stable DNA polymerase to amplify genes or gene segments for subsequent cloning and analysis:

1. The process begins with the denaturation of genomic DNA by raising the temperature to near the boiling point of water.
2. An excess of primers is then introduced, allowing them to anneal to the denatured DNA by lowering the temperature.
3. Each DNA strand is replicated as the DNA synthesis reaction proceeds near the optimal temperature for the thermostable DNA polymerase.
4. This cycle is repeated multiple times, leading to the exponential amplification of DNA. The amplified DNA bands can be visualized on agarose gels by staining with an intercalative fluorescent dye, such as ethidium bromide.

Throughout this process, each DNA strand is replicated by permitting the DNA synthesis reaction to occur at or near the temperature optimum of the thermostable DNA polymerase.

In practice, this cycle is repeated multiple times, resulting in exponential amplification of the DNA until the amplified DNA band becomes visible on agarose gels. Visualization is achieved by staining the DNA with an intercalative fluorescent dye such as ethidium bromide.

During each cycle, both DNA strands are replicated by permitting the DNA synthesis reaction to proceed at the optimal temperature for the thermostable DNA polymerase.

Complementarity and Hybridization

Molecular researchers utilize various forms of complementarity to differentiate and identify specific macromolecules of interest from a multitude of other molecules. Complementarity in this context refers to the sequence-specific or shape-specific molecular recognition that takes place when two molecules interact and form a binding connection. For instance, the two strands of a DNA double helix bind together due to their complementary sequences. Furthermore, an antibody can bind to a specific region of a protein molecule because their shapes complement each other, facilitating molecular recognition and interaction.

Probe-Target Complex

Complementarity between a probe molecule and a target molecule can result in the formation of a probe-target complex. This complex can be traced and located if the probe molecules are tagged with radioactivity or an enzyme. Locating this complex allows researchers to gather valuable information about the target molecule. In a solution, various types of hybrid molecular complexes, often referred to as hybrids, can coexist in a range of different combinations, providing researchers with diverse insights and analytical possibilities.

Probe-Target Hybridization Complex

In molecular biology, hybridization plays a crucial role in various interactions:

DNA–DNA: A single-stranded DNA (ssDNA) probe molecule can form a double-stranded, base-paired hybrid with a ssDNA target if the probe sequence is the reverse complement of the target sequence.

DNA–RNA: Similarly, a single-stranded DNA (ssDNA) probe molecule can form a double-stranded, base-paired hybrid with an RNA molecule, even though RNA is typically single-stranded, if the probe sequence is the reverse complement of the target RNA sequence.

Protein–Protein: In contrast to nucleic acid interactions, an antibody probe molecule can form a complex with a target protein molecule.

Hybridization possesses two crucial features:

1. **Specificity**: Hybridization reactions exhibit high specificity. The probe molecule will exclusively bind to targets with a complementary sequence or, in the case of antibodies, sites with the correct three-dimensional shape. This specificity ensures precise molecular recognition.
2. **Selectivity**: Hybridization reactions can occur even in the presence of large quantities of molecules that are similar but not identical to the target. This means that a probe can successfully locate and bind to one particular target molecule within a mixture containing a vast number of related molecules that lack complementary sequences or shapes. This selectivity is essential for the accuracy and sensitivity of molecular detection and analysis techniques.

These properties enable the use of hybridization in sensor technology.

Regarding DNA Damage

DNA damage can be categorized into two primary types:

- Endogenous damage results from factors within the organism, such as the attack by reactive oxygen species produced as byproducts of normal metabolic processes. This type includes spontaneous mutations and processes like oxidative deamination. It also encompasses replication errors.
- Exogenous damage, on the other hand, is caused by external agents. This includes factors like ultraviolet (UV) radiation from the sun, other types of radiation (e.g., X-rays and gamma rays), hydrolysis or thermal disruption, exposure to certain plant toxins, and mutagenic chemicals, particularly aromatic compounds that act as DNA intercalating agents, as well as viruses.
- DNA mutations are relatively rare occurrences in which the cell's DNA replication and repair processes fail, resulting in permanent changes in the DNA. These permanent changes are known as mutations and can have significant consequences. Even a mutation affecting a single nucleotide pair can severely impact an organism's fitness, particularly if it occurs in a crucial position within the DNA sequence.
- The structure and function of each protein are dependent on its specific amino acid sequence. Therefore, any alteration in this sequence can lead to a protein functioning poorly or not functioning at all. For instance, humans rely on the protein hemoglobin for oxygen transport in the blood. A permanent change in just one nucleotide can cause cells to produce hemoglobin with an incorrect amino acid sequence, leading to conditions like sickle-cell anemia.
- In sickle-cell anemia, a mutation results in the production of hemoglobin that is less soluble than normal hemoglobin. This abnormal hemoglobin forms fibrous precipitates, which give rise to the characteristic sickle shape of affected red blood cells.

Nucleic acids play a crucial role in biosensing, with a primary focus on leveraging their inherent property of sequence complementarity. These biorecognition elements, often used as probes, are meticulously designed to complement the specific sequence of the target molecule. One of their significant advantages lies in their capability to amplify the target DNA sequence through the polymerase chain reaction (PCR) and enhance the signal generated by the biosensor, greatly enhancing detection sensitivity.

Besides DNA, other nucleic acids that undergo biorecognition processes in biosensing include ribonucleic acid (RNA) and peptide nucleic acids (PNAs). PNAs are synthetic DNA analogs known for their superior hybridization characteristics. They offer improved chemical stability over a wide range of temperatures and pH levels and enhanced enzymatic stability as they are resistant to nucleases and proteases. These properties make PNAs a more feasible choice compared to DNA in various biosensing applications, contributing to the versatility and reliability of nucleic acid-based biosensors.

Affinity-Based Biorecognition

Affinity-based biosensors encompass a wide range of biorecognition elements, including antibodies, DNA, synthetic or natural receptors, and more. These biosensors consist of sensing element complexes that engage in crucial interactions between the receptor and the ligand through non-covalent binding involving various functional groups. Such affinity complexes can include antigen-antibody, DNA/oligonucleotides, or protein–protein interactions.

In recent years, immuno-complexes have garnered significant attention in research and the scientific community, particularly in the development of immunosensors. Antibody-based detection methods are considered the gold standard in clinical analysis. Antibodies are generated in response to antigens, such as toxins, chemicals, drugs, virus particles, bacterial toxins, and other foreign substances. While cell clone technology has increased antibody production, it still faces limitations related to pH and temperature sensitivity, short shelf life, and susceptibility to degradation. Challenges persist in terms of in situ measurements, regeneration and reuse, antibody generation for toxic compounds, and signal enhancement for small compounds.

However, amperometric methods have gained prominence due to their rapid and sensitive detection capabilities, leading to increased commercialization. They address a wide array of issues in fields such as health, environment, security, and biowarfare.

The underlying principle of affinity biosensors hinges on the formation of a stable and selective complex between the appropriate analyte and receptor element in a short timeframe, resulting in a measurable signal. Immunosensors, along with other types of biosensors, aim to fulfill critical requirements, including a strong biological foundation, operation in complex matrices, rapid response times, and miniaturization.

In recent years, advancements in biotechnology, nanotechnology, and surface chemistry have catalyzed breakthroughs in affinity biosensors. These innovations have led to the discovery of more advanced and highly sensitive affinity interactions, making use of biologically derived materials such as aptamers or phages. These developments hold promise for the continued enhancement of biosensor technology.

3.2.3 Antibodies: Structure, Binding Kinetics, and Principles

Antibodies are a class of proteins produced as part of the body's initial defense against foreign substances known as antigens (Ag). The binding between antigens and antibodies (Ag–Ab binding) relies on non-covalent interactions, demonstrating remarkable selectivity and sensitivity. Antigens are foreign substances, typically nanometer-sized, with a molecular weight exceeding 1.5 kilodaltons (kDa).

Antibodies are immunoglobulins, and they fall into five primary classes: IgG, IgM, IgA, IgD, and IgE. However, when it comes to immunosensors, only IgG antibodies are commonly used.

Here are the definitions for reference:

- Antigen (Ag): A molecule that triggers a specific immune response when introduced into an organism.
- Antibody (Ab): A glycoprotein produced in response to an antigen that exhibits specificity for the antigen and binds to it through non-covalent interactions.
- Antigens and antibodies exhibit highly specific binding to each other. This specific interaction is known as an antigen-antibody reaction, abbreviated as an Ag–Ab reaction. It forms the foundation for humoral immunity or antibody-mediated immunity and is instrumental in the detection of infectious diseases. The non-covalent interactions that underlie antigen-antibody binding encompass hydrogen bonds, ionic bonds, hydrophobic interactions, and van der Waals forces.
- Origin of Antibodies: Antibodies are produced by B cells, also known as B-lymphocytes, and are situated on the surface of these cells. Each B cell generates a unique antibody. Memory B cells circulate in the bloodstream and patrol the body's tissues to provide protection against potential future infections.

Antibody Structure and Function

Antibodies, also known as immunoglobulins, are crucial proteins in the immune system. Each antibody is composed of four polypeptide chains, consisting of two heavy chains and two light chains, which combine to form a distinctive "Y"-shaped molecule. The mechanism used by antibodies to eliminate antigens is determined by their constant region. Antibodies are categorized into five major classes: IgM, IgG, IgA, IgD, and IgE, based on the structure of their constant regions and their specific immune functions.

The variable region of antibodies is further divided into hypervariable (HV) and framework (FR) regions. HV regions directly interact with a portion of the antigen's surface and are sometimes referred to as complementarity-determining regions (CDRs). In an antigen-antibody reaction, the antibody binds to the antigen, and the part of the antigen that interacts with the antibody is called an epitope or antigenic determinant. The region of the antibody that recognizes the epitope is referred to as a paratope.

Antibodies carry out various functions aimed at eliminating the antigen that triggered their production. Some of these functions are not dependent on the specific class

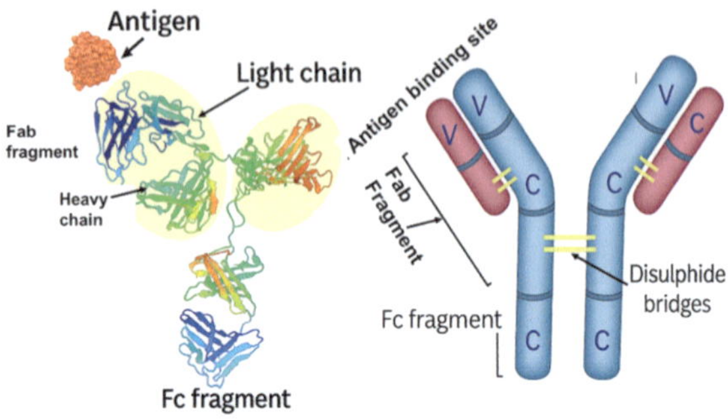

Fig. 3.11 Structure of an antibody comprises four polypeptide chains, which include two heavy chains and two light chains, all joined together to form a distinctive "Y"-shaped molecule. The antigen-binding sites, responsible for binding to antigens, are found at the fragment, antigen-binding (Fab) region of the antibody. When an antibody binds to an antigen, it forms an antigen-antibody complex, facilitating various immune responses and processes [6]

(isotype) of immunoglobulin and are determined by the antigen-binding capacity of the antibody, as defined by its variable and hypervariable (idiotypic) regions. For instance, one antibody may bind to a toxin, preventing it from entering host cells and exerting its biological effects. Similarly, a different antibody may bind to the surface of a virus, blocking its entry into host cells.

On the other hand, certain antibody functions are contingent upon the immunoglobulin class (isotype) to which the antibody belongs. These isotype-dependent functions play distinct roles in the immune response (Fig. 3.11).

Models for Antibody Binding: In the simplest reaction, a small univalent ligand binds reversibly to a specific site on the antibody. The antibody may have an identical sites, and when the ligand concentration is saturating, all these sites will be occupied. The number of bound ligand molecules per antibody molecule will then equal n, which is known as the antibody valence. (For example, it's 2 for IgD, IgG, and IgE, and 10 for IgM.)

If all the binding sites are identical and independent, the binding reaction can be represented as:

$$S + L \leftrightarrows \mathrm{SL}. \tag{3.4}$$

Here, S represents a binding site on the antibody, and L is the ligand. At equilibrium, the rates of the forward and reverse reactions are equal:

$$k_{\text{forward}}[S][L] = k_{\text{backward}}[\mathrm{SL}] \tag{3.5}$$

- [SL] represents the concentration of occupied antibody sites.
- $[S]$ represents the concentration of free sites.
- $[L]$ represents the concentration of unbound ligand.

The ratio of the rate constants for the forward and reverse reactions gives the association constant (K), which quantifies the affinity of the antibody for the ligand. K is typically measured in units of liters per mole (L/mol), which can also be expressed as M^{-1} (inverse molar units). This value reflects how strongly the antibody and ligand bind to each other. A higher K indicates a stronger binding affinity between the antibody and the ligand.

Equation (3.5) can be rearranged in terms of measure parameters by dividing both the numerator and denominator by the antibody concentration [Ab]. This results in:

$$K = \frac{[\mathrm{SL}]}{([\mathrm{Ab}] - [\mathrm{SL}])[L]}. \tag{3.6}$$

We can simplify this expression by redefining the number of ligand molecules bound per antibody molecule, [SL]/[Ab].

By introducing the variable "r" to represent the number of ligand molecules bound per antibody molecule, we can obtain the following expression:

$$K = \frac{r}{(n - r)[L]} \tag{3.7}$$

which is the same as:

$$r = Kn - Kr[L] \tag{3.8}$$

r the ratio of the concentration of bound ligand to the concentration of antibody molecules placed in the system,

n the number of ligand binding sites on the antibody molecule (i.e., its valence), and

L the concentration of unbound ("free") ligand.

Scatchard plot: In Scatchard plot analysis, the results are typically plotted on a graph of $r/[L]$ versus r. This type of plot should yield a straight line with a slope of $-1/Kd$, assuming that the initial assumption holds true that all the binding sites are identical and independent.

When the concentration of free ligand $[L]$ is very high, the ratio $r/[L]$ approaches 0, and r approaches n, which is the antibody valence. In other words, the number of bound ligand molecules per antibody molecule (r) equals the antibody valence (n) under these conditions. This behavior is an important characteristic observed in Scatchard plots.

Immunoassays: Conventional immunoassays are analytical tools that leverage antigen-antibody (Ag–Ab) complexes. The fundamental principle of immunoassays involves the detection of specific macromolecules known as antigens within a mixture of cells or molecules, effectively distinguishing between self and non-self-substances. Immunoassays are of great interest due to their remarkable capability for information processing and are commercially available in various formats for detecting very low levels, typically in the range of 10^{-12}–10^{-9} mol/L, of substances such as hormones, enzymes, viruses, toxins, tumor antigens, and bacterial antigens.

Commercially available immunoassays rely on a crucial feature of Ag–Ab complexes, as well as the detection of signals through various means. To facilitate this detection, a range of labels is utilized, including enzymes (enzyme immunoassays, EIA, or enzyme-linked immunosorbent assay, ELISA), radioisotopes (radioimmunoassay, RIA), and chemiluminescent compounds (chemiluminescent immunoassay, CLIA). Among these, ELISA is the most widely used and comes in two major formats: (1) competitive format and (2) sandwich format. ELISA is favored for its sensitivity, rapid detection, cost-effectiveness, simplicity, and suitability for use in field settings.

3.2.4 Phages/Viruses for Biorecognition

Bacteriophages (or phages), which are resilient alternatives to antibodies, are the most abundant biological entities found in various environments. They can be selectively pooled from a library to bind specifically to target molecules. Bacteriophages have a lytic or lysogenic cycle. The lytic cycle leads to the death of the host, whereas the lysogenic cycle leads to integration of phage into the host genome. Phages lack their own reproduction machinery, and they propagate by incorporating their DNA into host bacteria to create new virions, which then infect new hosts (lytic phages). Alternatively, they can utilize bacterial internal machinery to reproduce new phages or remain dormant after injecting their DNA into the host (lysogenic phage). Consequently, molecular engineering of phages has the potential to lead to the development of engineered phage bioprobes. Phages have gained prominence as potentially promising biorecognition elements in biosensing applications [7].

Molecular and genetic engineers now possess the capability to manipulate genetic material through phage display, enabling the preparation of phages as substitutes for antibodies in probe applications. Phage display stands as a powerful technology that harnesses viruses (phages) to multiply and propagate within host bacterial cells. For a comprehensive overview of the potential perspectives and applications of phage display technology, readers are encouraged to explore the in-depth review by Petrenko and Smith [8]. Figure 3.12 presents a schematic representation of both the lytic and lysogenic cycles involved in phage DNA propagation (https://thegeneticsofvirusesandbacteria.weebly.com/diagrams.html). Extensive research efforts spanning several decades have been dedicated to identifying phages capable of targeting various bacterial strains, including *Salmonella*, *Escherichia coli*, *Campylobacter*, *Bacillus anthracis*, *Listeria*, and others, by utilizing phage display technology. This research has resulted in significant advancements in the field of phage-based biosensors, where phages serve as highly specific biorecognition elements with a strong affinity for target bacteria [9, 10] (Fig. 3.13).

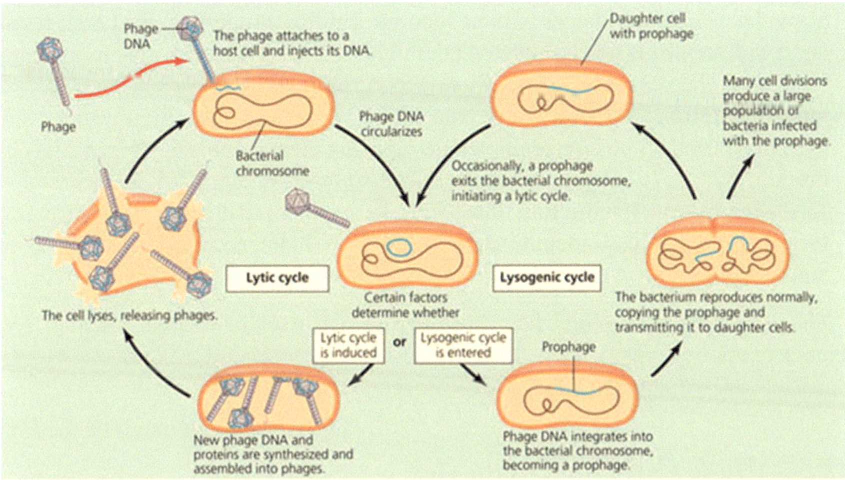

Fig. 3.12 **a** 3D view of bacteriophage. **b** Enterobacteria phage P4, which infects *E. coli*. It is commonly found in mammalian intestines and sewage. **c** A group of phages, shown in green, attacking an *E. coli* cell by injecting their DNA through the cell membrane. https://prints.sciencesource.com/art/photographs/bacteriophages

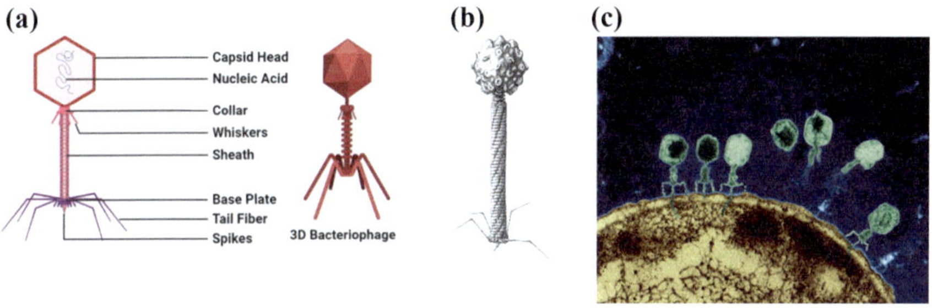

Fig. 3.13 Illustration of the bacteriophage replication pathways: the lytic and lysogenic cycles. The process begins when a phage attaches to a bacterial host and injects its DNA. The phage DNA then circularizes, leading to two possible pathways. In the lytic cycle, the phage commanders the host machinery to replicate its DNA and synthesize proteins, forming new phages. Eventually, the host cell lyses, releasing the newly formed phages. In the lysogenic cycle, the phage DNA integrates into the bacterial chromosome as a prophage. The host cell reproduces normally, passing the prophage to daughter cells. Under certain conditions, the prophage may exit the chromosome and enter the lytic cycle

Here are some key features and benefits of phage display:

- **Excellent Substitutes for Antibodies**: Phages can serve as excellent substitutes for antibodies in the detection of various important analytes.

- **Superior Stability**: Phages exhibit superior stability properties, including resistance to extreme pH levels, high temperatures (above 80 °C), and organic solvents, surpassing the stability of antibodies.
- **High Host/Target Affinity and Specificity**: Phages display high affinity and specificity for their host/target molecules, making them reliable tools for biorecognition.
- **Ease of Synthetic Production**: Phages can be synthesized through a cost-effective fermentation process, making their production more accessible compared to antibodies.

Phage display technology has proven to be an asset in various research and diagnostic applications.

3.2.5 Whole Cells and Organelles

In nature, both whole cells and subcellular organelles possess remarkable sensing capabilities, allowing them to adapt and thrive in diverse environments. Organelles, which function like organs within cells, have specific roles such as storing genetic information in nuclei and producing chemical energy in mitochondria. Genetic engineering techniques have advanced the immobilization of both whole cells and organelles onto transducers, enhancing their sensitivity and selectivity for analyte detection.

The rapid progress in synthetic biology has significantly expedited the development of biosensors based on whole cells and organelles, particularly for clinical applications. These biosensors employ bacterial or mammalian cells, including laboratory strains and commensal bacteria from the human microbiota, as depicted in Fig. 3.14.

These biosensors can be considered "natural biosensors" because they operate on the principle of "biology sensing biology." They possess several key properties:

1. *Robustness*: Cells exhibit robustness and can function effectively in harsh environments. They act as a buffering system to maintain the functionality of sensor components, such as receptors and transcription factors, even in extreme conditions such as low pH, high temperature, or osmotic stress.
2. *Sensitivity*: Biosensors are designed to detect molecules within a biologically relevant concentration range, making them highly sensitive to specific analytes.
3. *Selectivity*: Sensor components possess highly specific recognition sites that enable them to detect analytes at the molecular level and distinguish between isomers. Signal transduction mechanisms are often orthogonal, allowing for precise discrimination of similar biological signals in complex environments.
4. *Evolvability*: Sensor properties can be improved through directed evolution, offering an advantage over chemical systems that are limited to screening approaches.

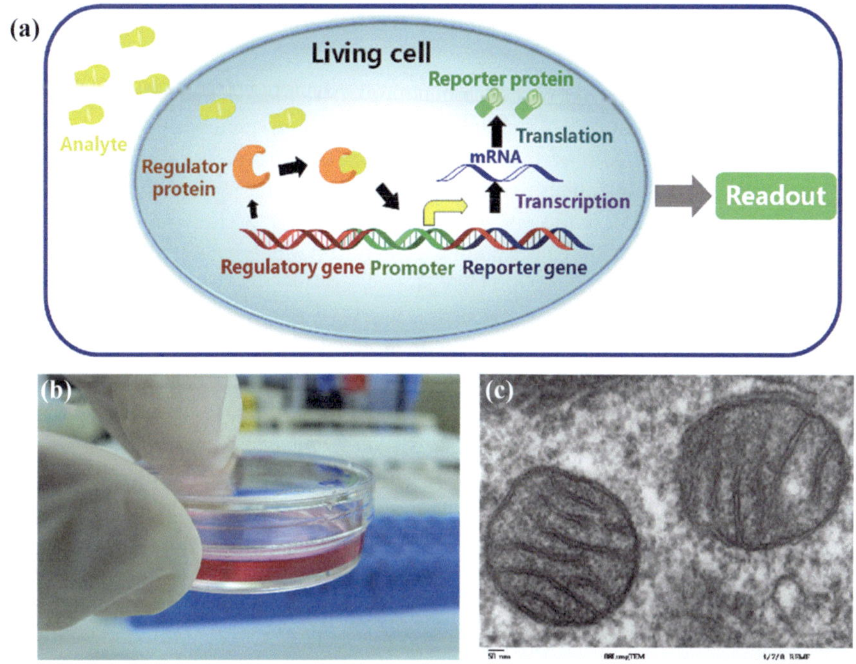

Fig. 3.14 **a** A schematic representation of a whole-cell-based biosensor; **b** a depiction of cell culture; **c** a scanning electron microscopy (SEM) image of mitochondria cells. Please note that image **a** is sourced from MDPI, while images **b** and **c** are from Wikipedia (no permission needed) [11]

5. *Online/Continuous Monitoring*: Cells continuously monitor their environment, eliminating the need for interrupted or periodic sampling.

These properties collectively make whole cells and organelles-based biosensors promising tools for a wide range of applications, particularly in clinical and environmental monitoring.

3.3 Biosynthetic Biorecognition Elements

With the advancements in recombinant technology and the availability of numerous expression systems, it is now feasible to produce substantial quantities of receptor proteins. These receptor proteins, often referred to as biosynthetic biorecognition elements, play a crucial role in addressing the challenges that may arise from the lack of suitable recognition elements.

3.3.1 Aptamers (DNA/Protein), RNA Aptamers

Aptamers are miniature nucleic acids, either single-stranded DNA (ssDNA) or RNA, that are selected from a library of randomly sequenced oligonucleotides. They possess the remarkable ability to bind to a wide range of target proteins with high affinities and specificities. Aptamers were first discovered through an in vitro selection process called Systematic Evolution of Ligands by EXponential enrichment (SELEX) (Fig. 3.15). This breakthrough was achieved independently by both Gold's group and Szostak's group in 1990 [12, 13].

Since their discovery, extensive research has been conducted to select aptamers against various targets, ranging from small molecules to proteins and even whole cells. Aptamers can be single-stranded oligonucleotides, either ribonucleic acid (RNA) or deoxyribonucleic acid (DNA), as well as peptide molecules. They exhibit high affinity and specificity for their target molecules and undergo changes in sequence, three-dimensional structure, and folding pattern upon binding to the target.

Over the past decade, aptamers have emerged as a compelling alternative to antibodies in bioanalytical systems. They offer significant advantages in biosensing, serving as specific antagonists and biomarkers for therapeutics, pharmaceuticals, and clinical applications. Aptamers are known for their high affinity, sensitivity, specificity, simplicity in structural modification, ease of production, and regenerability. These properties have positioned aptamers as formidable competitors to antibodies, particularly in diagnostic and biosensing applications.

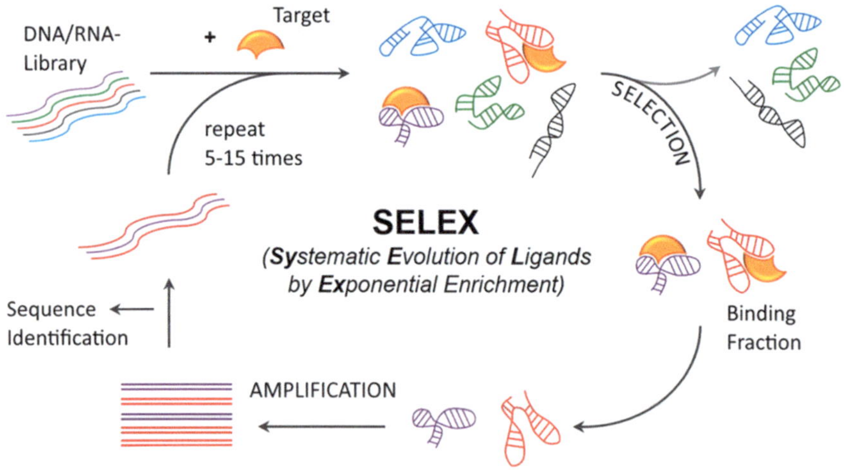

Fig. 3.15 Systematic Evolution of Ligands by EXponential (SELEX) enrichment process. A random nucleic acid library undergoes incubation with a target molecule, followed by the separation of unbound molecules from bound ones. The bound nucleic acids are subsequently eluted, amplified through polymerase chain reaction (PCR), and become an enriched library for the next cycle. For each target, 6–12 consecutive cycles are executed, resulting in the final enriched library, which is subsequently cloned and sequenced [14]

Multiple aptamers have been successfully selected against a wide array of targets, showcasing their versatility and potential applications. Here are a few examples of the types of targets aptamers have been developed for:

1. Metal ions: Aptamers have been designed to selectively bind to metal ions such as potassium (K^+), mercury (Hg^{2+}), and lead (Pb^{2+}).
2. Small organic molecules: Aptamers can recognize and bind to various small organic molecules, including amino acids, adenosine triphosphate (ATP), antibiotics, vitamins, and even substances like cocaine.
3. Organic dyes, peptides, and proteins: Aptamers have been selected to specifically interact with organic dyes, peptides, and proteins. Examples include aptamers that target proteins like thrombin, growth factors, and HIV-associated peptides.
4. Whole cells or microorganisms: Aptamers have been developed to bind to entire cells or microorganisms, including bacteria. This capability is particularly valuable in the field of microbial detection.

The diversity of aptamer targets highlights their potential in a wide range of applications, including diagnostics, anti-bioterrorism efforts, and environmental and food analysis. The availability of aptamers tailored to different targets opens up possibilities for innovative bioassay tools and technologies across various domains.

3.3.2 Phages

Additionally, phages have also gained prominence as potentially promising biorecognition elements in biosensing applications. Phage display is a powerful technology that harnesses viruses (phages) to multiply and propagate within host bacterial cells. This method offers several advantages in biosensing and biorecognition research.

Here are some key features and benefits of phage display:

- **Excellent Substitutes for Antibodies**: Phages can serve as excellent substitutes for antibodies in the detection of various important analytes.
- **Superior Stability**: Phages exhibit superior stability properties, including resistance to extreme pH levels, high temperatures (above 80 °C), and organic solvents, surpassing the stability of antibodies.
- **High Host/Target Affinity and Specificity**: Phages display high affinity and specificity for their host/target molecules, making them reliable tools for biorecognition.
- **Ease of Synthetic Production**: Phages can be synthesized through a cost-effective fermentation process, making their production more accessible compared to antibodies.

Phage display technology has proven to be an asset in various research and diagnostic applications due to these advantageous features (Fig. 3.16).

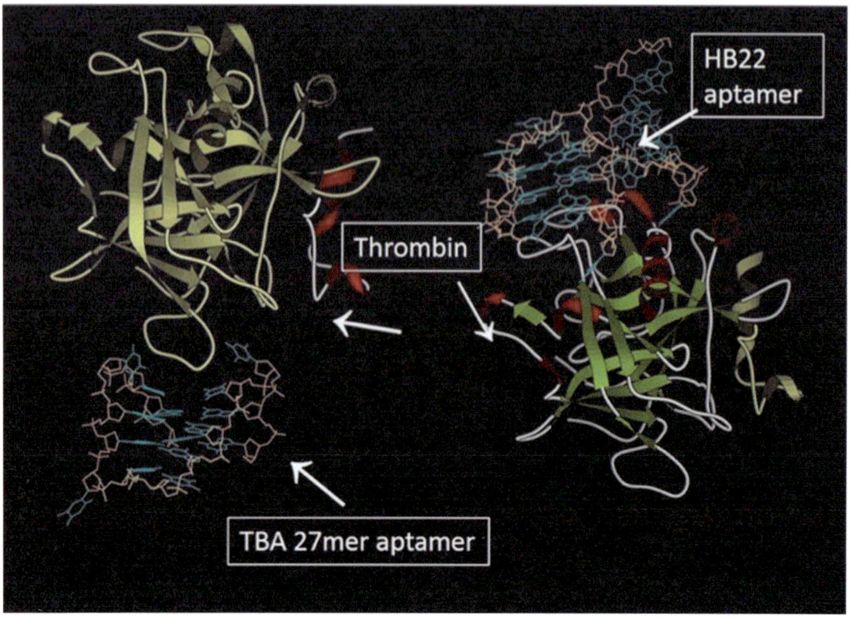

Fig. 3.16 Molecular structures of thrombin bound to two distinct aptamers. The TBA 27-mer aptamer (bottom left) and the HB22 aptamer (top right) are shown interacting with thrombin protein domains (center). The ribbon diagrams highlight the protein's secondary structure (β-sheets and α-helices), while the aptamers are rendered in stick models to indicate their nucleotide composition and binding conformation. This figure illustrates the specific and high-affinity recognition of thrombin by aptamers, a key feature in aptamer-based biosensing and therapeutic applications [15, 16]

3.3.3 DNA Chips, CRISPR/Cas Systems

DNA chips comprise DNA-based microarrays where DNA probes are deposited in numerous microspots on a solid support. DNA microarrays serve the purpose of measuring the expression levels of numerous genes simultaneously or genotyping multiple regions of a genome. Within each DNA spot, there exist picomoles (10–12 mol) of a particular DNA sequence, referred to as probes, reporters, or oligos. These sequences can represent short sections of genes or other DNA elements used for hybridization with a cDNA or cRNA sample (known as the target) under high-stringency conditions. Typically, the hybridization between probes and targets is detected and quantified by identifying fluorophore-, silver-, or chemiluminescence-labeled targets, allowing for the determination of the relative abundance of nucleic acid sequences in the target.

Another innovative approach to developing sensitive and selective nucleic acid-based biosensors involves CRISPR (Clustered Regularly Interspaced Short Palindromic Repeats) and CRISPR-associated proteins (Cas), collectively known as CRISPR/Cas systems. These systems function as nucleic acid-based adaptive immune systems that provide protection to microorganisms against viral infections.

CRISPR/Cas systems are based on the remarkable ability of bacteria to store fragments of foreign phage genomes in CRISPR loci, essentially as a form of memory from past encounters. These CRISPR loci, along with surrounding repeats, serve as guides for Cas endonucleases to selectively recognize sequences in foreign genomes and combat invading phages. In addition to their significant biological impact, CRISPR/Cas systems have garnered considerable attention in the field of biosensing [17, 18].

The CRISPR array consists of a series of DNA repetitions, typically 25–35 base pairs (bp) in length, interspersed with spacer sequences of around 30–40 bp. These spacer sequences effectively retain the memory of previous invasions and are passed on to the microorganism's progeny during cell division. The defense mechanism can be divided into three main stages: adaptation, CRISPR RNA (crRNA) biogenesis, and target interference.

- **Adaptation Phase**: This phase is initiated when the microorganism encounters an invader for the first time. During this phase, the CRISPR loci encode a Cas enzyme that identifies and incorporates a portion of the foreign genetic material, known as a protospacer, into a spacer within the CRISPR array, effectively recording the invasion event.
- **Biogenesis Phase**: In this phase, a crRNA guide is synthesized using the stored protospacer sequence. The crRNA guide is then employed by the Cas endonuclease enzyme to form a ribonucleoprotein (RNP) interference complex. This complex can recognize and specifically cleave the nucleic acid of the invading phage during the interference or "immunity" phase, which is the third phase.

Programmable Cas enzymes are considered potent biorecognition elements for detecting nucleic acids. They enhance the sensitivity, specificity, cost-effectiveness, and speed of isothermal amplification-based molecular diagnostic laboratory techniques. Furthermore, they have been successfully integrated into point-of-care (POC) devices, exhibiting performance levels comparable to routine techniques, as demonstrated in Fig. 3.17.

3.3.4 Recombinant DNA Technology: Genetically Modified Cells

Before delving into the concept of recombinant DNA technology, it's essential to address the fundamental question: "What is recombinant DNA?" Recombinant DNA, often referred to as "chimera," is the term used to describe the amalgamation of one DNA fragment with another. The term "recombinant" signifies this combining process. Scientists can create a new strand of DNA by merging two or more distinct DNA strands. Typically, this process involves merging the DNA from two different organisms. There are three primary methods for obtaining recombinant DNA (rDNA): transformation, phage introduction, and non-bacterial transformation.

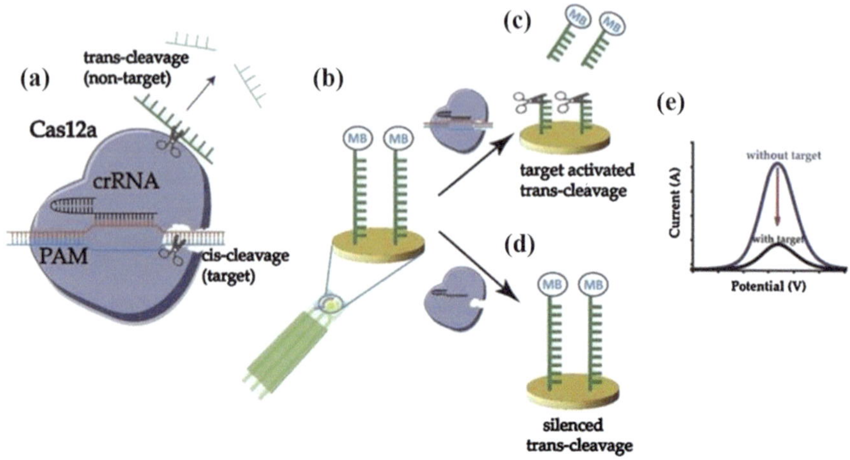

Fig. 3.17 Schematic representation of a CRISPR-Cas12a-based biosensing mechanism. **a** Cas12a-crRNA complex binds to the protospacer adjacent motif (PAM) on the target DNA, initiating cis-cleavage of the target and trans-cleavage of nearby non-target DNA. **b** Sensor platform with molecular beacons (MBs) attached to the electrode surface. **c** Presence of the target activates Cas12a-mediated trans-cleavage, cutting the MBs. **d** In the absence of the target, MBs remain intact, resulting in silenced trans-cleavage. **e** Electrochemical signal change: the current response (A) versus potential (V) graph shows a significant shift between the "with target" and "without target" conditions, indicating successful target detection [19]

1. Transformation: The initial step in the transformation process involves selecting a DNA fragment for insertion into a vector. Next, this DNA fragment is cleaved using a restriction enzyme, followed by ligation, where the cleaved DNA is inserted into the vector using DNA ligase. The inserted DNA usually carries a selectable marker that aids in identifying recombinant molecules. Frequently, an antibiotic resistance marker is utilized. In this scenario, a host cell lacking a vector will perish when exposed to a specific antibiotic, whereas the host containing the vector will survive due to its antibiotic resistance. The vector is then introduced into a host cell through a process called transformation. For instance, *Escherichia coli* (*E. coli*) can serve as a possible host cell. The host cells must be suitably prepared to uptake the foreign DNA.

Selectable markers can encompass antibiotic resistance, alterations in color, or any other characteristic that distinguishes transformed hosts from untransformed ones. Different vectors possess varying properties, rendering them suitable for diverse applications. These properties may include symmetrical cloning sites, size, and high copy number.

2. Phage Introduction: Phage introduction, also known as transfection, is a process similar to transformation, but instead of bacteria, a phage is employed. In this method, in vitro packaging of a vector is utilized. Lambda or MI3 phages are commonly used to generate phage plaques containing recombinant DNA. These

recombinants can be discerned from non-recombinants through various selection techniques.

3. Non-bacterial Transformation: Non-bacterial transformation closely resembles the process of transformation, with the only distinction being the absence of bacteria, such as *E. coli*, as the host.

Recombinant DNA plays a pivotal role in the development of genetically modified cell-based biosensors (GCB). GCBs are sensing methodologies that employ living microbial cells, whether prokaryotic or eukaryotic, to provide information about the chemical composition of their surrounding environment. These organisms serve as indicators of the environmental conditions to which they are exposed.

3.3.5 Site-Directed Mutagenesis, Directed Evolution, and Domain Swapping

Site-directed mutagenesis is a PCR-based technique used to introduce specific nucleotide mutations into a sequence within a plasmid vector. This method enables the investigation of the relative importance of particular amino acids in terms of protein structure and function. Typical mutations are designed to disrupt or map protein–protein interactions, mimic or block posttranslational modifications, or suppress enzymatic activity.

Similarly, directed protein evolution is an alternative approach recently introduced to protein engineering, which was honored with the Nobel Prize in Chemistry in 2018. Predicting the impact of individual amino acid substitutions on protein properties is often challenging, but it is easier to measure this effect experimentally. The method involves generating a diverse set of protein sequences, some of which may represent potential solutions to the engineering objective. These proteins are then screened for desirable properties and functions. In directed evolution, sequence diversification and screening are typically iterated multiple times, with additional amino acid substitutions accumulating in each round. Each round brings the protein sequence closer to the target protein engineering goal. Directed evolution makes use of molecular biology techniques for sequence diversification and methods from biochemistry, analytical chemistry, and microbiology to screen the resulting proteins for desired properties.

Domain swapping can be described as the exchange of structural components between two or more identical protein monomers. This structural phenomenon plays a significant role in the mechanism of oligomerization for certain proteins. In oligomeric structures, the monomer units become intricately intertwined with one another. Engineered domain swapping has the potential to generate self-assembling materials that preserve and combine the functions of the parent proteins or introduce new functionalities, such as linking on/off switching properties with selective and sensitive sensing capabilities. Domain swapping has emerged as a potent method for creating a diverse range of novel bioreceptors.

3.3.6 Biomimetic Systems: Molecular Imprinted Polymers

Molecularly imprinted polymers (MIPs) are synthetic polymer materials with various biomimetic functions, including molecular recognition, catalytic activity, and stimuli-responsive behavior. The preparation of MIPs involves the formation of complexes between a template molecule (which can be the target molecule or its derivative) and functional monomers. These interactions can be either covalent or non-covalent. Subsequently, co-polymerization is carried out in the presence of a cross-linker. Once the polymerization is complete, the template molecule is removed, resulting in the formation of binding cavities within the polymer. These cavities are complementary in terms of size, shape, and functional group arrangement to the template molecule (see Fig. 3.18).

The combination of molecularly imprinted polymers (MIPs) with various types of transducers results in a synergistic device. MIPs exhibit resilience to pH variations, organic environments, and changes in ionic strength, making them highly suitable for sensor development. This technology is particularly advantageous for the detection of non-electroactive molecules, which are substances that cannot undergo electrochemical reactions, such as pesticides and drugs. These non-electroactive molecules can be quantified using affinity techniques or catalytic secondary molecule assays.

However, secondary molecule assays have certain drawbacks, including potential interactions between the secondary molecule and the target molecule, solvent-related

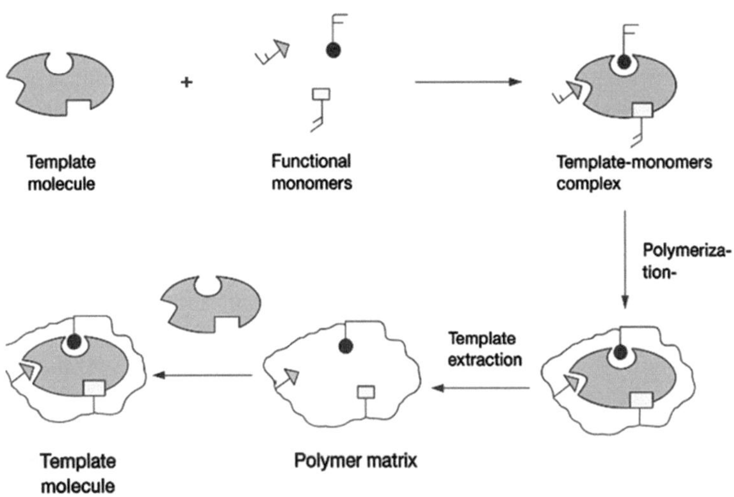

Fig. 3.18 Schematic illustration of the molecular imprinting process. Functional monomers interact with a template molecule to form a pre-polymerization complex. Polymerization around this complex forms a rigid polymer matrix. After polymerization, the template molecule is extracted, leaving behind specific recognition sites complementary in shape and functional groups to the template. The resulting molecularly imprinted polymer (MIP) can selectively rebind the target molecule, enabling applications in sensing, separation, and drug delivery [20]

issues (such as the possibility of template and secondary molecule not being soluble in the same solvent), and sensitivity to harsh reaction conditions that may affect the measurement of the target molecule. Therefore, affinity-based measurements are preferred for the detection of non-electroactive molecules.

Traditional biocompounds like DNA, antibodies, proteins, and receptors are often used to develop affinity-based systems. However, these biomolecules tend to be expensive, challenging to immobilize onto transducers, and difficult to optimize under specific conditions. To overcome these limitations, MIP-based sensors have been developed using various types of transducers, including electrochemical, optical, and piezoelectric devices. This approach offers a more cost-effective and practical solution for the detection of non-electroactive molecules.

References

1. M, S., FH White, J. & CB, A. Reductive cleavage of disulfide bridges in ribonuclease. *Science (80-).* **125**, 691–692 (1957).
2. Bodanszky, M. *Principles of Peptide Synthesis.* (Springer Science & Business Media, 2012).
3. Eisenberg, D. The discovery of the α-helix and β-sheet, the principal structural features of proteins. *Proc. Natl. Acad. Sci. U. S. A.* **100**, 11207–11210 (2003).
4. Schulz, A. R. *Enzyme Kinetics: From Diastase to Multi-Enzyme Systems.* (Cambridge University Press, 1994).
5. Berg, J. M.; Tymoczko, J. L.; Gatto, G. J.; Stryer, L. *Biochemistry.* (W.H. Freeman & Company, New York, 2015).
6. Nelson DL, Cox MM, L. A. *Lehninger Principles of Biochemistry.* (WH Freeman, New York, 2021).
7. Sh. Balasubramanian, I. Sorokulova, V. Vodyanoy, A. Simonian. Lytic phage as a specific and selective probe for detection of *Staphylococcus aureus* - A surface Plasmon resonance spectroscopic study. Biosensors and Bioelectronics 22 (2007) 948–955.
8. Chem. Rev. 1997, 97, 2, 391–410.
9. Al-Hindi, R. R. *et al.* Bacteriophage-Based Biosensors: A Platform for Detection of Foodborne Bacterial Pathogens from Food and Environment. *Biosensors* **12**, 1–22 (2022).
10. Li, S. *et al.* Direct detection of *Salmonella typhimurium* on fresh produce using phage-based magnetoelastic biosensors. *Biosens. Bioelectron.* **26**, 1313–1319 (2010).
11. Gui, Q., Lawson, T., Shan, S., Yan, L. & Liu, Y. The application of whole cell-based biosensors for use in environmental analysis and in medical diagnostics. *Sensors (Switzerland)* **17**, 1–17 (2017).
12. Tuerk, C. & Gold, L. Systematic Evolution of Ligands by Exponential Enrichment: RNA Ligands to Bacteriophage T4 DNA Polymerase. *Science (80-).* **249**, 505–510 (1990).
13. Ellington, A., Szostak, J. In vitro selection of RNA molecules that bind specific ligands. *Nature* **346**, 818–822 (1990).
14. Kruspe, S. & Giangrande, P. H. Aptamer-siRNA chimeras: Discovery, progress, and future prospects. *Biomedicines* **5**, 1–20 (2017).
15. Irene, Russo Krauss, Andrea, P., Merlino, A., Mazzarella, L. & Sica, F. Duplex–quadruplex motifs in a peculiar structural organization cooperatively contribute to thrombin binding of a DNA aptamer. *Acta Crystallogr. D. Biol. Crystallogr.* **69**, 2403–2411 (2013).
16. Russo Krauss, Irene; Merlino, Antonello; Randazzo, Antonio; Novellino, Ettore; Mazzarella, Lelio; Sica, F. High-resolution structures of two complexes between thrombin and thrombin-binding aptamer shed light on the role of cations in the aptamer inhibitory activity. *Nucleic Acids Res.* **40**, 8119–8128 (2012).

17. Hille, F. *et al.* The Biology of CRISPR-Cas: Backward and Forward. *Cell* **172**, 1239–1259 (2018).
18. Zhou, L., & Simonian, A. L. CRISPR/Cas Technology: The Unique Synthetic Biology Genome-Editing Tool Shifting the Paradigm in Viral Diagnostics, Defense, and Therapeutics. *Annu. Rev. Biomed. Eng.* (2024) https://doi.org/10.1146/annurev-bioeng-081723-013033.
19. Dai, Y. *et al.* Exploring the Trans-Cleavage Activity of CRISPR-Cas12a (cpf1) for the Development of a Universal Electrochemical Biosensor. *Angew. Chemie - Int. Ed.* **58**, 17399–17405 (2019).
20. Vasapollo, G. *et al.* Molecularly Imprinted Polymers: Present and Future Prospective. *Int J Mol Sci* 12, 5908–5945 (2011).
21. Magnetoelastic Biosentinels for the Capture and Detection of Low-Concentration Pathogens in Liquid, International Journal on Smart Sensing and Intelligent Systems 7(5):1-4,https://doi.org/10.21307/ijssis-2019-075

Chapter 4
Immobilization and Surface Chemistry

4.1 Immobilization: Fundamentals and Methods

The immobilization of proteins onto solid supports is essential for achieving high surface density, enabling the use of small sample volumes and minimizing non-specific protein adsorption—thus improving detection performance. However, preserving the protein's native conformation and biological activity during immobilization remains a major challenge.

Protein immobilization offers numerous advantages, such as enabling multiple uses of a single enzyme batch, rapid termination of reactions by removing the enzyme or the substrate, enhanced stability, prolonged half-life, predictable activity decay, and the elimination of repeated reagent preparation.

Various methods can be employed to immobilize enzymes onto solid supports, including physical adsorption, entrapment, encapsulation, cross-linking, layer-by-layer (LbL) assembly, covalent binding, and bioaffinity.

4.2 Physical Adsorption

Physical adsorption is a straightforward and rapid method for constructing biosensors. However, it often leads to random enzyme orientations, potentially compromising enzymatic activity. This method relies on weak, reversible interactions—such as van der Waals forces, ionic interactions, hydrogen bonding, and occasionally hydrophobic forces—between the enzymes or cells and the support material. Despite the relative weakness of these forces, their cumulative effect can facilitate stable adsorption. In electrochemical biosensors, for example, proteins with positively charged amino acid residues can be electrostatically adsorbed onto negatively charged surfaces by immersing the electrode in the protein solution.

A. Simonian et al., *Biosensors as Analytical Tools for the 21st Century*,
https://doi.org/10.1007/978-3-032-01726-0_4

Physical adsorption is a simple and rapid method for manufacturing biosensors. While it offers speed and simplicity, it has the drawback of often resulting in unfavorable enzyme orientations that can reduce their functionality. This immobilization method relies on reversible surface interactions between enzymes/cells and support materials, primarily driven by electrostatic forces such as van der Waals forces, ionic interactions, hydrogen bonding, and occasionally hydrophobic bonding. Although these forces are relatively weak, they occur in sufficient numbers to facilitate binding. In the case of electrochemical sensors, proteins in solution with positively charged amino acid residues can be electrostatically adsorbed onto a negatively charged surface by simply immersing the electrode in the solution.

4.3 Entrapment

Entrapment involves immobilizing biomolecules within a matrix or gel lattice, rather than on its surface. The porosity of the lattice is optimized to retain the biomolecules while allowing free diffusion of substrates and products. While the matrix may hinder mass transport and affect kinetics, it also offers protective benefits—shielding immobilized enzymes or cells from harmful interactions with external molecules.

The entrapment method differs from physical adsorption in that it involves immobilizing biomolecules within a lattice structure, such as a gel, rather than allowing them to remain free in solution. The porosity of the gel lattice is carefully controlled to strike a balance between preventing the leakage of proteins or cells while allowing the free movement of substrates and products. It's important to note that the support material acts as a barrier to mass transfer, which can affect reaction kinetics. However, this barrier also has its advantages as it prevents harmful cells, proteins, and enzymes from interacting with the immobilized biomolecules.

4.4 Encapsulation

Encapsulation of biomolecules involves enveloping biological components within semi-permeable membranes. This process is similar to entrapment, as it restricts the enzymes/cells in a confined space. Large proteins or enzymes cannot pass into or out of the capsule, but small substrates and products can freely cross the semi-permeable membrane. Various materials have been used to create microcapsules, typically ranging from 10 to 100 μm in diameter. Popular materials for this purpose include nylon and cellulose nitrate.

4.5 Cross-Linking

This type of immobilization is support-free and involves connecting cells (or enzymes) to each other, forming large, three dimensional complex structures. Cross-linking can be achieved through chemical or physical methods. Chemical cross-linking usually entails the formation of covalent bonds between cells using a bi- or multifunctional reagent, such as glutaraldehyde or toluene diisocyanate. However, the toxicity of these reagents limits their application to living cells and many enzymes.

4.6 Layer-by-Layer (LbL)

The layer-by-layer technique, initially demonstrated by Decher et al. and Hong involves the sequential deposition of molecular multilayers with tailored architecture through consecutive electrostatic interactions between cationic and anionic polyelectrolytes. This concept allows for structurally well-defined materials connected by weak electrostatic interactions, which is particularly valuable for immobilizing enzymes while preserving their native structure and activity.

While other interactions such as hydrogen bonding, biospecific recognition, coordination, hydrophobic interactions, and dipole–dipole interactions have been studied, the assembly of multilayers involving the adsorption of charged species on a solid surface through electrostatic interactions is widely used in the LbL methodology. In some cases, LbL construction requires the presence of functional groups, such as thiols, disulfides, amines, acids, or silanes, to create a self-assembled monolayer (SAM).

The construction of LbL structures involves sequential deposition on a charged solid substrate, such as a negatively charged surface immersed in a solution containing a cationic polyelectrolyte, resulting in the formation of a layer of polycationic species. The relatively high concentration of species in the solution leads to excess adsorption, effectively reversing the surface charge. The layer is then rinsed and immersed in a solution containing an anionic polyelectrolyte until a new layer is adsorbed, restoring the original surface charge. Careful selection of polyelectrolytes and the number of layers allows for precise control of the properties of these polyelectrolyte multilayers (PEM). An initial example of an LbL multilayer film involved poly(styrene sulfonate) (PSS) as the polyanion and poly(allylamine hydrochloride) (PAH) as the polycation is shown in Fig. 4.1.

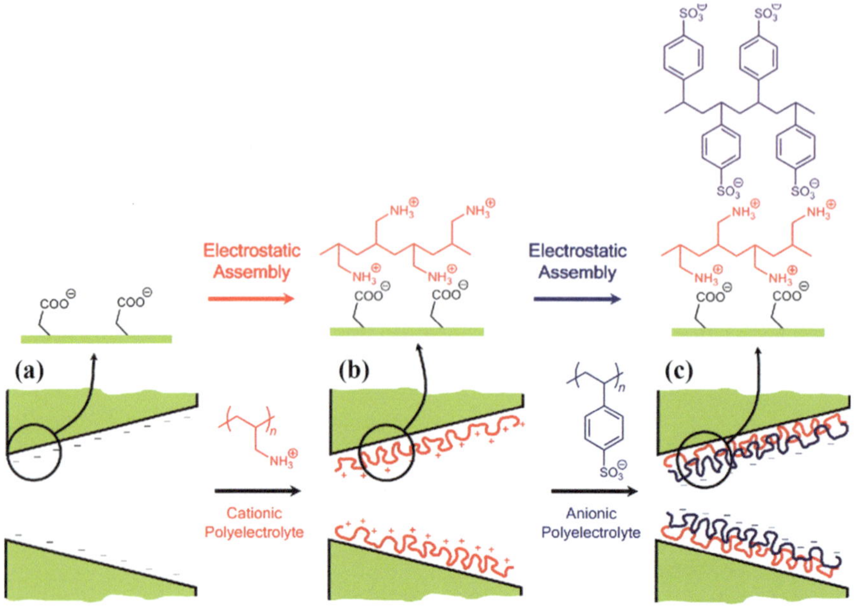

Fig. 4.1 Schematic illustration of electrostatic layer-by-layer assembly of polyelectrolytes on a charged substrate. **a** Initial negatively charged surface (carboxylate groups). **b** Adsorption of a cationic polyelectrolyte (e.g., poly(allylamine)) onto the negatively charged surface via electrostatic attraction. **c** Subsequent adsorption of an anionic polyelectrolyte (e.g., sulfonated polystyrene) onto the cationic layer, forming a multilayer structure. The zoomed-in cross sections show surface morphology changes during each deposition step, highlighting the buildup of polyelectrolyte layers through alternating charge interactions [1]

4.7 Covalent Binding

One of the most intensely studied immobilization techniques involves the formation of covalent bonds between enzymes and the support matrix. The covalent binding method is based on the principle of attaching enzymes to water-insoluble carriers through covalent bonds. This technique requires the formation of covalent bonds between the enzyme/cell and a support material. Typically, these bonds form between functional groups present on the surface of the support and functional groups belonging to amino acid residues on the surface of the enzyme. Numerous amino acid functional groups are suitable for participating in covalent bond formation.

Amine coupling: Enzymes or other protein molecules can be immobilized by coupling them to the surface using ionizable side chains found in most significant amino acids. Examples of such amino acids include aspartic acid (–COOH group), glutamic acid (–COOH group), and lysine (–NH_2 group). Amine coupling is a versatile method and is generally applicable to most biomacromolecules containing amine groups. This process is also referred to as amination. Lysine residues on the exterior of the protein surface are commonly used as anchors for coupling. However, in some cases, the abundance of lysine residues can increase heterogeneity and limit conformational flexibility, as they attach at multiple points on the surface. This method may also be less suitable for proteins with acidic ligands (pI < 3.5), ligands where an amine is in the active site of the biomolecule, or molecules possessing many amine groups. To perform amine coupling, proteins need to be dissolved in a low ionic-strength buffer and passed over the activated surface with common functional groups present on cross-linking or modification reagents.

The most commonly used agent for coupling with amine groups, forming stable amide bonds, is N-Hydroxysuccinimide (NHS). NHS esters are reactive groups formed through carbodiimide activation of carboxylate molecules. NHS ester-activated cross-linkers and labeling compounds react with primary amines under slightly alkaline conditions (pH 7.2–9), resulting in stable amide bonds and releasing N-hydroxysuccinimide (NHS). NHS esters then react with nucleophilic groups of the proteins, creating strong amide bonds (Fig. 4.2). Immobilization efficiency depends on several parameters, including pH, concentration, ionic strength, and reaction time. However, these conditions need to be optimized for each type of protein.

Thiol coupling: The choice of thiol (–SH) coupling mostly depends on the availability of thiol groups on the ligand. However, it is relatively easy to introduce thiol groups to the ligand. Thiol chemistry is more robust than amine chemistry, so the coupling conditions are less critical. Thiol coupling cannot be used under strong reducing conditions, as the disulfide bond is unstable in such conditions. The amino acid cysteine contains the thiol group, capable of forming internal disulfide bonds that contribute to the stability of the protein. Since cysteines are not as abundant as lysines, random immobilization is less likely to occur.

Maleimide: Thiols' reaction with maleimides is a widely used method for bioconjugation and labeling of proteins and peptides due to their high selectivity. The reaction involves an addition reaction between the maleimide double bond and thiol groups to form stable thioether bonds. The reaction can be carried out specifically in a pH range of 6.5–7.5, while at higher pH values, some cross-reactivity with –NH_2 has been observed. In aqueous conditions, maleimide reacts rapidly but undergoes slow hydrolysis, which may cause problems in protein handling. Homobifunctional (bis(maleimidohexane) (BMH)) and heterobifunctional (N-(-maleimidocaproyl)succinimide (EMCS) and sulfosuccinimidyl

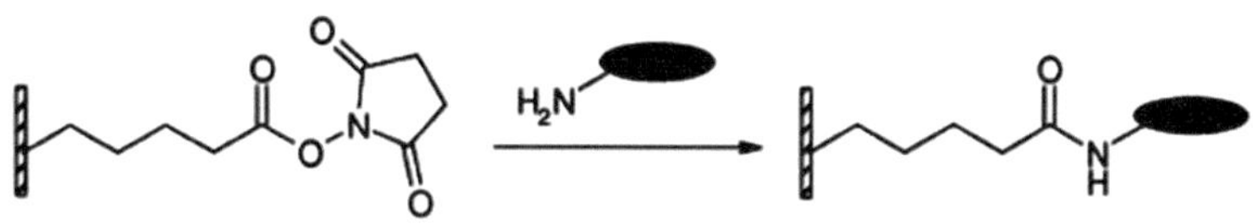

Fig. 4.2 Amide bond formation through NHS ester chemistry for surface immobilization. A surface functionalized with an N-hydroxysuccinimide (NHS) ester reacts with a primary amine-containing molecule to form a stable amide bond, resulting in the covalent attachment of the molecule to the surface

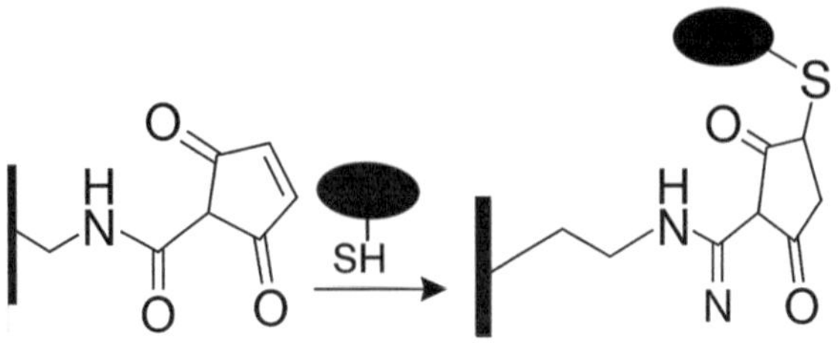

Fig. 4.3 Schematic representation of the covalent immobilization of a thiol-containing molecule onto a maleimide-functionalized surface via a thioether bond. The reaction proceeds through Michael addition, resulting in the stable conjugation of the thiol to the maleimide moiety

4[p-maleimidophenyl] butyrate (sulfo-SMPB)) linkers have been developed for maleimide coupling (Fig. 4.3).

Carboxylic acid coupling: It might be interesting to immobilize proteins via carboxyl groups since aspartic and glutamic acid usually constitute the major fraction of surface groups on proteins. This method is an alternative to adsorbing proteins via multipoint interactions onto aminated slides. These mild covalent immobilizations are achieved using carbodiimide chemistry. The method for activating –COOH was proposed by Fernandez-Lafuente et al. using β-galactosidase from *Aspergillus oryzae* as a model protein [2].

Carbodiimide: Water-soluble carbodiimide hydrochloride is very useful for amide formation and protein modification. 1-Ethyl-3-(3-dimethylaminopropyl)-carbodiimide (EDC) is a water-soluble derivative of carbodiimide. Carbodiimide catalyzes the formation of amide bonds between carboxylic acids or phosphates and amines by activating carboxyl or phosphate groups. EDC reacts with a carboxyl group and forms an amine-reactive intermediate that is unstable in aqueous solutions. The reagent can be used to make ether links from alcohol groups and ester links from acid and alcohols or phenols, and peptide bonds from acid and amines. The 5′-end phosphorylated DNA is activated in a reaction between the phosphate group and EDC. The reaction between the EDC-activated phosphate group in the 5′-end of the DNA and 1-methylimidazole forms a more stable compound. The reactive group on the NucleoLink surface indicated with an X reacts with the phosphate group on the DNA coupled to 1-methylimidazole, creating covalent binding. Water and 1-methylimidazole are by-products of the reaction. N-Hydroxysuccinimide (NHS) is often used to assist the carbodiimide coupling in the presence of EDC. The reaction involves the formation of the intermediate active ester (the product of condensation of the carboxylic group and N-hydroxysuccinimide), which further reacts with the amine function to yield the amide bond. NHS reacts with carboxyl groups to form NHS ester derivatives, which may be considered the most common functional groups for creating reactive acylating agents (Fig. 4.4). NHS esters were first introduced as reactive ends of homobifunctional cross-linker [3, 4]. Today, the majority of commercially available amine-reactive cross-linking or modification reagents utilize NHS esters.

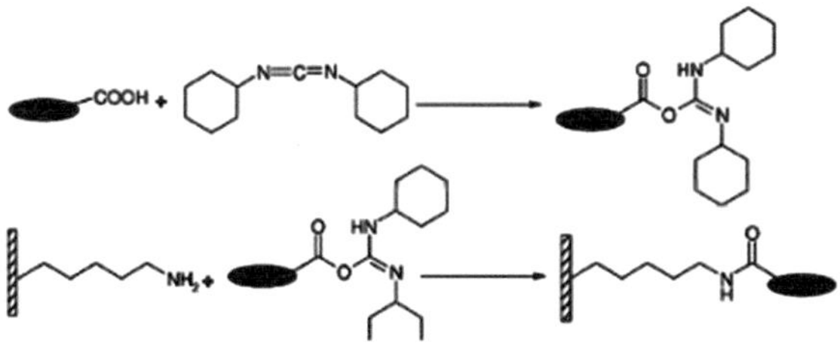

Fig. 4.4 Two-step reaction scheme for covalent immobilization using a diisocyanate linker. In the first step, a carboxyl-functionalized molecule reacts with hexamethylene diisocyanate to form a reactive isocyanate intermediate. In the second step, the resulting intermediate reacts with an amine-functionalized surface to form a stable urea linkage, resulting in the covalent attachment of the molecule to the surface

Aldehyde: Aldehyde coupling occurs with polysaccharides and glycoconjugates, which have cis-diols and sialic acid groups that are relatively easy to oxidize to aldehydes. The aldehyde surface is suitable for coupling numerous biomolecules. The amine groups of proteins can be easily attached to the surface with a support consisting of aldehyde groups. The chemistry is called reductive amination, which involves the formation of an initial Schiff base between the aldehyde and amine groups, which is then reduced to a secondary amine by the addition of sodium cyanoborohydride ($NaCNBH_3$). During the coupling process, the reducing agent, cyanoborohydride, is mild enough not to cleave disulfides in most proteins and not to reduce the aldehyde reactants—only the Schiff base intermediates. This coupling can achieve an immobilization efficiency of up to 85%, depending on the type and amount of ligand present (Fig. 4.5).

Isothiocyanate: The reaction of an aromatic amine with thiophosgene ($CSCl_2$) yields isothiocyanate (–NCS), which forms a stable bond with primary amine groups. The reaction between the proteins and the thioisocyanate group involves unprotonated amines. Hence, as a result of lower pKa, alpha-amino acids would react faster at a neutral pH than x-amino groups, allowing, in principle, selective coupling of proteins through the formation of stable bonds mainly with alpha-amino groups. Isothiocyanate groups have higher stability and relative ease of preparation.

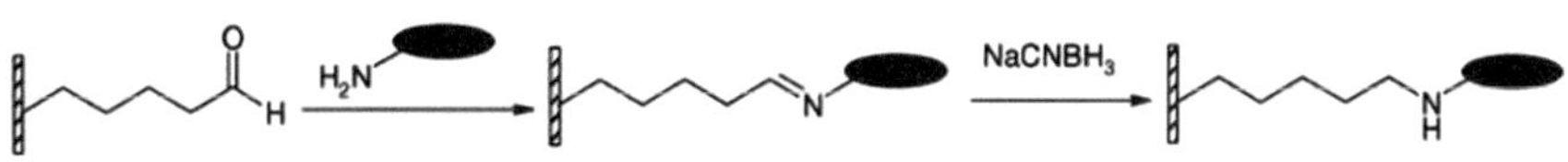

Fig. 4.5 Schematic representation of reductive amination for covalent immobilization. An aldehyde-functionalized surface reacts with a primary amine-containing molecule to form a Schiff base (imine intermediate), which is subsequently reduced by sodium cyanoborohydride ($NaCNBH_3$) to yield a stable secondary amine linkage, resulting in covalent attachment of the molecule to the surface

Epoxide: Epoxy chemistry has many advantages and seems to be an appropriate system for immobilizing proteins due to its stability at neutral pH values, wet conditions, and reactivity with several nucleophilic groups to form strong bonds with minimal chemical modification of the protein. However, these interactions have been limited by their extremely slow reaction rate, although some studies have shown that they can react at a high rate with nearby epoxy groups in the same support. A two-step fast-paced mechanism has been hypothesized, promoted by an increased number of epoxy supports on the surface, which has been hypothesized to improve the efficiency of immobilization and stability of the immobilized proteins. Epoxy-amino group supports coated with a layer of ethylenediamine promote physical adsorption via amine groups, and then covalent linkage via epoxy groups. A new support with conventional epoxy has shown a faster immobilization rate for Lipase from *Candida rugosa* and β-galactosidase from *A. oryzae*, which demonstrated an increased half-life of immobilized glutaryl acylase, invertase, and β-galactosidase from *Thermus* sp. by 3–5 times. Epoxy chemistry with thiol has been proposed to immobilize proteins via thiol-disulfide exchange followed by covalent immobilization via epoxy chemistry.

Glutaraldehyde: Glutaraldehyde is the most popular cross-linker in use. The reaction with proteins and other amine-contained molecules is expected to proceed through the formation of Schiff bases. Schiff bases alone would not yield stable cross-linked products without reduction. Cross-linking using GA polymers is difficult to reproduce. To utilize glutaraldehyde, dissolve the protein or other amine-containing macromolecule to be modified at a concentration of 1–10 mg/mL in a buffer with a pH between 7 and 10. The higher the pH, the more efficiently Schiff base formation will occur. Phosphate, borate, and carbonate buffers at concentrations ranging from 0.1 to 1 M are acceptable. Avoid amine-containing buffers such as Tris or glycine, as they will react with glutaraldehyde. Add a quantity of glutaraldehyde equal to a tenfold molar excess over the number of amines to be modified. A typical concentration of glutaraldehyde in the reaction mixture is 1.25%. In some cases, trial experiments may be necessary to check the solubility of the resultant modified protein. Scale back the quantity of glutaraldehyde added if precipitation occurs. React for at least 2 h at 4 °C. Quickly isolate the modified protein by gel filtration using Sephadex G-25 or an equivalent method.

Silica surface: Silica surfaces can arise from the dehydration of hydrated silica preparations ("silicic acids"), where water elimination occurs as the temperature rises and condensation of silanol functions Si–OH produces siloxane bridges Si–O–Si. Rehydration of the surfaces can occur upon sintering, causing the silica network to rearrange and compact. On the other hand, silanones > Si=O surfaces are produced by any fracture in a silica piece, regardless of its amorphous or crystalline structure. In the presence of water, all these structures are subjected to hydration processes leading to the formation of silanol groups. Therefore, the commonly observed silica surfaces exhibit two main chemical functions: silanols (Si–OH) and siloxanes (Si–O–Si). These silanols are significantly more acidic than their alcohol counterparts, rendering silica surfaces negatively charged within a wide range of pH values. Moreover, they show a remarkable tendency to act as donors in hydrogen bonding.

Bioaffinity: Affinity immobilization is achieved by exploiting the specificity of enzymes for their support under different physiological conditions. This affinity can occur when the matrix is pre-coupled to an affinity ligand for the target enzyme, or when the enzyme is conjugated to an entity that develops an affinity toward the matrix. Affinity adsorbents and complex affinity supports have been used to host higher amounts of enzymes, leading to increased stability and efficiency. Bioaffinity layering is an improvisation of this technique that exponentially increases enzyme-binding capacity and reusability due to the presence of non-covalent forces, such as coulombic interactions.

Avidin-biotin: A more controllable and better reproducible method for immobilizing biomolecules is possible using the high-affinity, bioselective interaction between biotin and avidin. Avidin, which is a tetrameric protein, binds four biotin molecules with a dissociation constant of about 10^{-15} M. Thus, it represents one of the strongest non-covalent interactions known. The choice of streptavidin-biotin coupling follows when amine or thiol coupling is unsatisfactory or unsuitable. Nucleic acids, polysaccharides, and glycoconjugates can be relatively easily biotinylated using a variety of reagents and functional groups. Acidic and acid-sensitive proteins can be coupled at neutral pH with high efficiency.

Tag: The polyhistidine tag, a genetically encoded affinity tag, is the most well-known. It typically consists of six sequential histidine residues and chelates transition metals, including Cu(II), Co(II), Zn(II), or Ni(II). Ni(II) is the most commonly employed metal. Typically, a support bearing a chelating moiety, such as nitrilotriacetic acid or imino-diacetic acid, is treated with a solution of the relevant metal salt to produce a support presenting the metal ions. This metal-activated support is then used for protein immobilization through chelation with the histidine residues of the tag.

Another straightforward method of protein immobilization involves the use of enzymatically active fusion proteins. In this approach, a protein of interest is fused to an enzyme (capture protein) that selectively reacts with an immobilized substrate analogue or inhibitor. This enables the formation of a covalent bond between the enzyme and the substrate on the surface. For example, a study by Martinez reported the use of the serine esterase cutinase from the filamentous fungus [5].

Endogenous fusion proteins: These proteins have been identified to bind to metal ions, thus competing with the desired histidine-fused protein. For most applications, the desired histidine-fusion protein must be purified before use. The strength of the binding interaction is also relatively weak (Kd $\approx 1^{-10}$ μM). However, proteins bearing tags with 10^{-12}. His residues or two separate histidine tags have been shown to provide improvements of up to 1 order of magnitude, enabling in situ immobilization of the target protein.

References

1. Azzaroni, O. & Lau, K. H. A. Layer-by-layer assemblies in nanoporous templates: Nano-organized design and applications of soft nanotechnology. *Soft Matter* **7**, 8709–8724 (2011).
2. Torres, R. *et al.* A novel heterofunctional epoxy-amino sepabeads for a new enzyme immobilization protocol: Immobilization-stabilization of β-galactosidase from *Aspergillus oryzae. Biotechnol. Prog.* **19**, 1056–1060 (2003).
3. Lomant, A. J., & Fairbanks, G. Chemical probes of extended biological structures: synthesis and properties of the cleavable protein cross-linking reagent [35S] dithiobis (succinimidyl propionate). *J. Mol. Biol.* **104**, 243–261 (1976).
4. Bragg, P. D., & Hou, C. Subunit composition, function, and spatial arrangement in the Ca^{2+}- and Mg^{2+}-activated adenosine triphosphatases of *Escherichia coli* and *Salmonella typhimurium. Arch. Biochem. Biophys.* **167**, 311–321 (1975).
5. Chrislaine, M., Pieter, D. G., Marc, L., Gaston, M. & Christian, C. *Fusarium solani* cutinase is a lipolytic enzyme with a catalytic serine accessible to solvent. *Nature* **356**, 615–618 (1992).

Chapter 5
Novel Materials for Biosensing

Over the past two decades, the incorporation of novel nanomaterials in biosensor development has opened new research frontiers. These materials offer unique properties—such as substrate-specific responsiveness, nanoscale dimensions, and tailored chemical functionality—which enable them to mimic or even replace traditional biorecognition elements like redox enzymes, antibodies, and nucleic acids. Such materials are often termed "biomimetic" due to their ability to integrate the functional roles of nanomaterials and bioreceptors.

Nanotechnology is the study and application of materials at the nanoscale. Nanomaterial-based biosensors, which combine material science, molecular engineering, chemistry, and biotechnology, have demonstrated enhanced sensitivity and selectivity for detecting biomolecules. These systems can interact with or manipulate individual atoms and molecules, enabling advanced applications in biomolecular recognition, pathogen detection, and environmental monitoring. Nanotechnology products are generally categorized based on their composition—such as metal-based or carbon-based nanomaterials—and their distinctive physical properties, including conductivity and optical behavior.

5.1 Nanomaterials: Nanoparticles (Gold, Silver, Magnetic)

Among the wide variety of nanomaterials used in the development of biosensors, metal nanoparticles have been extensively studied due to their characteristic optoelectronic and thermal properties. It is worth noting that their unique optoelectronic properties, such as localized surface plasmon resonance (LSPR) and scattering, arise from the size of the material within the colloidal suspension, which differs significantly from the properties of the same material in bulk form. For instance, when comparing colloidal gold nanoparticles (average diameter = 50 nm) to bulk gold, they exhibit distinct optoelectronic properties because these phenomena occur on

A. Simonian et al., *Biosensors as Analytical Tools for the 21st Century*,
https://doi.org/10.1007/978-3-032-01726-0_5

different dimensional scales. The distinctive properties of these nanoparticles are attributed to localized surface plasmon resonance (LSPR), a nanoscale phenomenon responsible for sharp spectral absorption and scattering peaks, as well as strong electromagnetic near-field enhancements.

Before delving into specific types of nanoparticles (e.g., gold nanoparticles, silver nanoparticles, magnetic nanoparticles, etc.), it is beneficial to review the optoelectronic and magnetic properties of nanosized materials.

Optoelectronic Properties of Metal-Based Nanoparticles

When considering metal-based nanoparticles, the interactions between light and nanosized matter, such as absorbed light and scattering, are primarily related to the behavior of valence electrons. Specifically, the phenomenon of light absorption (which is visible as color to the naked eye) is associated with the collective oscillations of valence electrons, also known as plasmons. However, at the nanoscale, these oscillations depend not only on the material itself but also on the dimensions of the objects interacting with electromagnetic radiation, such as white light or lasers with fixed wavelengths. Additionally, the surrounding environment in which nanoparticles are immersed plays a crucial role.

When electromagnetic radiation strikes nanoparticles, it induces an instantaneous electric displacement along the field lines of the valence electrons, which, in a metal, can be considered as nearly free to move. Consequently, a net negative charge is generated at one end of the nanoparticle's surface, while an equivalent positive charge, due to ions, accumulates at the opposite end, as illustrated in Fig. 5.1.

This charge shift induces a pulling force on the electrons. However, the intensity of this pulling force depends on the surface charge density, which is influenced by both the surface shape and dimensions of the nanoparticle. It also relies on the surrounding

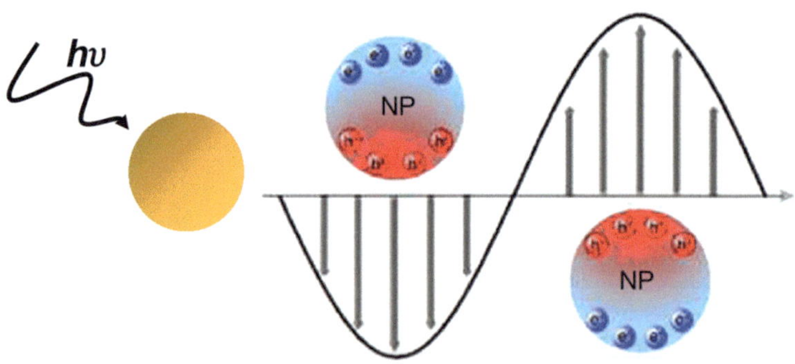

Fig. 5.1 Illustration of localized surface plasmon resonance (LSPR) in a metallic nanoparticle (NP). Upon illumination with light of energy $h\nu$, the conduction electrons on the nanoparticle surface collectively oscillate in response to the electric field of the incident light. These oscillations, represented along the waveform, lead to alternating charge distributions (positive and negative poles) on the NP surface, generating strong electromagnetic fields. This plasmonic behavior underpins the optical sensitivity of metallic nanoparticles in biosensing and spectroscopy applications

environment, particularly the solvent, as each solvent possesses a specific dielectric constant.

These collective excitations of the electron charges are referred to as localized surface plasmon resonances (LSPRs) to emphasize that these collective motions occur in the vicinity of the material's surface and are localized in a confined volume, notably within the nanoparticle. The generation of LSPR only occurs when the wavelength of light is significantly greater than the particle's diameter. Considering the wavelength of visible light (a few hundred nanometers), metal particles should exhibit a diameter in the nanometer range; otherwise, we would not observe any plasmonic effect (plasmon resonance). Plasmon resonance occurs only when the electromagnetic radiation's resonant frequency matches that of the nanoparticles' oscillation frequency. This interaction is extremely intense.

As the dimensions of nanoparticles increase, the absorbed frequency decreases, resulting in a red shift. Even more sensitive is the dependence of the absorbed frequency on the refractive index (dielectric constant) of the medium. A higher absorption constant leads to a decrease in the plasmonic frequency (and, consequently, in the absorption frequency) with an increasing red shift.

The interaction between light and nanoparticles is relatively simple to describe theoretically when the particles are spherical. Such a theory was first proposed in 1908 by Mie [1]. The Mie theory is a comprehensive mathematical and physical theory of the scattering of electromagnetic waves by homogeneous spherical particles. Unlike Rayleigh scattering, the Mie theory encompasses all possible ratios of particle radius to wavelength. It relies on the exact solution to Maxwell's equations for a spherical, homogeneous obstacle with plane wave excitation. Within the framework of the Mie theory, the scattered field is represented as an infinite series of fields of spherical multipoles, which are orthogonal at the sphere's surface. Mie's solution plays a crucial role in understanding light scattering by clouds of particles.

Thus, according to this theory, particles with a diameter of approximately 10 nm, when illuminated by visible light, almost entirely absorb radiation. However, larger particles, with diameters around 100 nm, scatter a significant portion of the radiation. The combination of absorbed and scattered radiation is referred to as the extinction coefficient, and it varies with frequency.

As previously discussed, at the nanoscale, the shape of objects strongly influences the frequencies at which incident light is absorbed. The same holds true for the colors of colloidal solutions containing nanoparticles. For example, it is possible to synthesize hollow nanoparticles with absorbed frequencies lower than those of similar but solid nanoparticles. This phenomenon arises from the interaction between the oscillating charges on the inner and outer surfaces of the particles. In the case of elongated particles, such as nanocylinders, two primary oscillation modes can be observed: one along the main axis at a lower frequency and another along the transverse axis at a higher frequency. Similarly, when one or more nanoparticles come into close proximity to each other, a common occurrence in colloidal solutions due to aggregation, they may develop shared electron oscillation modes. In optical terms, many aggregated nanoparticles will collectively appear as a larger nanoparticle, exhibiting a lower absorbed frequency.

A crucial characteristic of nanoparticles is their ability to enhance the electric field of incident radiation in the vicinity of their surfaces. Smaller bending radii result in more intense field enhancement. This enhancement can be explained by the concentration of electrons near the tips of the nanoparticles, generating a very high electric field in the immediate vicinity. This, in turn, enables the enhancement of various processes dependent on radiation. For example, when a molecule is in close proximity to a nanoparticle, its response to the electromagnetic field can be significantly amplified.

Gold nanoparticles: Beyond the theoretical understanding of the optoelectronic properties of metal-based nanoparticles, it's important to highlight their historical significance in art and sculpture, dating back to before the fourth century AD, primarily due to their distinctive optical properties. One of the most famous examples is the Lycurgus cup depicted in Fig. 5.2.

The Lycurgus Cup, a unique historical artifact, is an example of dichroic glass that changes color based on light exposure. This opaque green cup transforms into a translucent glowing red when illuminated from within. Upon analysis, it was discovered that the glass contains a very small quantity of tiny metal crystals, approximately 70 nm in size, composed of both Ag and Au in an approximate molar ratio of 14:1. It is the presence of these nanocrystals that gives the Lycurgus Cup its distinctive color-changing properties.

Until the Middle Ages, soluble gold was primarily known for its reputed therapeutic powers, particularly in treating conditions such as heart and venereal diseases.

(a) (b)

Fig. 5.2 Photographs of the famous Lycurgus cup which displays a different color depending on whether it is illuminated externally (**a**) or internally (**b**). https://www.tumblr.com/britishmuseum/120689869617/the-lycurgus-cup

The first book on colloidal gold was published in 1618 by the philosopher and medical doctor Francisci Antonii [2]. This book detailed various procedures for producing gold nanoparticles and their practical applications in medicine. Since then, numerous scientists have reported different methods for preparing gold nanoparticles, with a focus on wet chemical synthesis. In more recent times, Michael Faraday's notable publication reported the formation of deep red solutions of colloidal gold by reducing an aqueous solution of chloroaurate ($AuCl_4^-$) with phosphorus in CS_2 (a two-phase system). Faraday also explored the optical properties of thin films made from dried colloidal solutions, observing reversible color changes in response to mechanical compression (from bluish-purple to green). Following this pioneering work, thousands of scientific papers have been published on the synthesis, modification, properties, and assembly of metal nanoparticles, using a wide variety of solvents and substrates.

Industrial production of nanomaterials was primarily developed in the twentieth century. For instance, nanoparticles of carbon black (tire soot) have been used in the manufacturing of rubber tires for automobiles since the early twentieth century. Pigments such as SiO_2 and TiO_2 have been prepared through high-temperature combustion methods. Since the 1970s, the innovative development of nanoparticles has been driven by a combination of theory and experiments across various fields, including physics, chemistry, materials science, and biosciences. Specific phenomena, beyond optical properties, have led to new possibilities in various fields. Applications of nanoparticles in diverse areas require cost-effective and straightforward processes for synthesizing high-quality shaped nanoparticles.

Techniques for producing gold nanoparticles (AuNPs) have been reported using various methods, which can be classified as chemical, physical, and thermal methods.

Chemical methods: The synthesis of AuNPs through chemical reduction typically involves two main steps: (1) reduction by agents, such as borohydrides, aminoboranes, formaldehyde, hydrazine, hydroxylamine, polyols, citric and oxalic acids, sugars, hydrogen peroxide, carbon monoxide, sulfites, hydrogen, acetylene, and non-electronic reducing agents including electron-rich transition-metal sandwich complexes; (2) stabilization using agents, including trisodium citrate dihydrate, sulfur ligands (especially thiolates), phosphorus ligands, oxygen-based ligands, nitrogen-based ligands (including heterocyclic compounds), dendrimers, polymers, and surfactants (particularly cetyltrimethylammonium bromide (CTAB)). To prevent particle aggregation, some form of stabilizing agent is usually added.

Turkevich method: One of the most well-known techniques for synthesizing AuNPs is based on the reduction of $HAuCl_4$ by citrate in water, as first devised by Turkevich in 1951. In this method, the $HAuCl_4$ solution is boiled, and trisodium citrate dihydrate is rapidly added to the boiling solution while vigorously stirring. Within a few minutes, the solution's color changes from light yellow to wine red. This method produces AuNPs with an approximate diameter of 20 nm. Citrate ions in this technique serve a dual role as both stabilizing and reducing agents. The synthetic pathway is depicted in Fig. 5.3.

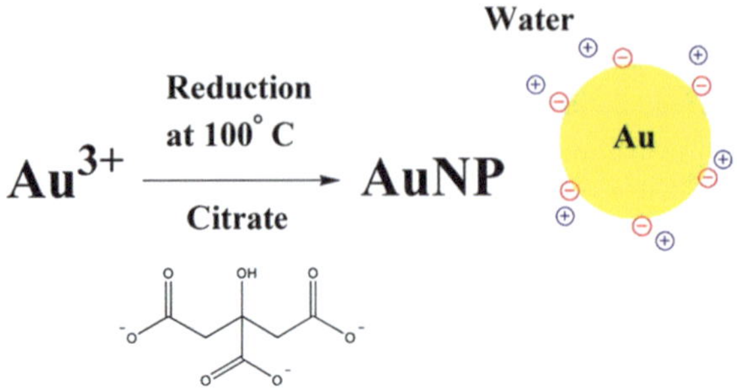

Fig. 5.3 Schematic representation of citrate-mediated synthesis of gold nanoparticles (AuNPs). Gold ions (Au^{3+}) are reduced by trisodium citrate at 100 °C, resulting in the formation of colloidal AuNPs. Citrate acts both as a reducing and stabilizing agent, capping the nanoparticles and providing electrostatic stabilization in aqueous solution through negatively charged surface groups that interact with the surrounding water molecules [3]

Turkevich method: This method was modified in 1973 by Frens, allowing the production of nanoparticles with different diameters in the range of 15–150 nm. This could be achieved by controlling the ratio of the reducing agent to the stabilizing agent (trisodium citrate/gold). Furthermore, this method has been further modified by several research groups to investigate the effect of citrate concentration on the aggregation/stabilization of colloidal gold, as well as considering both pH as an important factor and the order of reactants.

Brust–Schiffrin method: Another chemical method was discovered by Brust and Schiffrin in 1994 [4]. This method enables an easy approach to the synthesis of thermally stable and air-stable AuNPs with controlled size and low dispersity. In this technique, $AuCl_4^-$ is transferred to a toluene phase from an aqueous solution using tetraoctylammonium bromide (TOAB) as the phase-transfer agent and reduced by $NaBH_4$ in the presence of dodecanethiol. The addition of the reducing agent causes a color change in the organic phase, from orange to deep brown, clearly indicating the formation of AuNPs.

Green synthesis methods: The production of AuNPs through green synthesis is simple and facile. These NPs can be produced under moderate conditions without requiring elevated temperatures or pressures. The most common production method is the extracellular nanoparticle production method. The production is initiated with tetrachloroaurate salt ($HAuCl_4$). After mixing, the mixture is stirred to produce NPs. A color change to red or purplish indicates the successful production of AuNPs. This is followed by centrifugation and drying of the NPs for characterization and use. Notably, the simplicity of the production process varies depending on the choice of green materials. Green materials, such as bacteria, fungi, and enzymes, require delicate preparation steps. First, these organisms need to be cultured or propagated to obtain sufficient starting materials for the reaction. Additional steps may be required, such as ultrasonication and/or the use of enzyme separation processes to obtain purified green materials.

Green materials can act as both reducing and stabilizing agents for the synthesis of AuNPs. In brief, biomolecules from green materials initially reduce gold (III) ions to zero-valent gold. Then, NP stabilization, mediated by the biomolecules, occurs by covering the outer surface of the NPs through agglomeration. For example, grape pomace was utilized by González-Ballesteros et al. for the production of AuNPs due to its antioxidant properties [5]. Other types of berries that are rich in antioxidant content have also been chosen for the synthesis of AuNPs. Secondly, polar and water-soluble biocompounds are preferred for the reaction process. Gan et al. extracted various compounds from palm oil mill effluent using different solvents and showed that flavonoids, proteins, reducing sugars, and water-soluble alkaloids present in water extracts were all capable of reducing AuNPs [6]. Besides, small amounts of AuNPs were formed using an ethyl acetate extract containing polyphenols and alkaloids, while no reaction occurred using a hexane extract containing nonpolar compounds. The stability of the green-synthesized AuNPs is relatively sufficient for various applications.

Laser ablation: Laser ablation or photoablation is the process of removing material from a solid (or occasionally liquid) surface by irradiating it with a laser beam. At low laser flux, the material is heated by the absorbed laser energy and evaporates or sublimates. At high laser flux, the material is typically converted into a plasma. Usually, laser ablation refers to removing material with a pulsed laser, but it is possible to ablate material with a continuous wave laser beam if the laser intensity is high enough. This method is employed to produce metal-based nanoparticles by using a sacrificial surface dipped in a solution of a stabilizer like polyvinylpyrrolidone (PVP), etc.

Thermal methods: The microwave-assisted method of AuNPs synthesis is gaining popularity due to its high reaction rate, energy transfer efficiency, simplicity, product uniformity, and environmental friendliness. In this process, the material is directly heated by radiation instead of indirectly heated by thermal sources, leading to higher temperature homogeneity in the reaction mixture. During the heating process, microwaves interact with the dipole moments of the molecules present in the mixture, leading to their reorientation and reorganization to form AuNPs.

5.2 Nanocarbons: Carbon Nanotubes, Graphene, Nanowires

Iijima first discovered that multiwalled carbon nanotubes and single-walled nanotubes have fascinating properties that contribute to various applications ranging from polymers, energy storage, sensor materials, electronics to medicine and the environment. Prior to Iijima's discovery, in 1952, Radushkevich and Lukyanovich published clear images of 50-nm diameter tubes made of carbon in the Russian Journal of Physical Chemistry [7]. This discovery largely went unnoticed because the article was published in the Russian language. The long history of CNTs has been largely devoted to the investigation of their physical and chemical properties, as well as their growth mechanism.

Types of CNTs: The structure of CNTs can be classified according to their number of concentric graphene cylinders. A graphene cylinder has a similar structure to graphite but is a single isolated sheet of carbon.

Single-walled: A single-walled carbon nanotube (SWCNT) can be considered as formed by the rolling of a single layer of graphite (called a graphene layer) into a seamless cylinder (a long wrapped graphene sheet). SWCNTs consist of a single graphene cylinder with a diameter of 0.4 nm, although they are usually produced with a random distribution of diameters (0.7–2 nm) and exist in hexagonally close-packed bundles held together by van der Waals forces. SWNTs consist of two separate regions, a sidewall of the tube and the end cap of the tube, which contribute to different physical and chemical properties. They are excellent conductors because they exhibit important electrical properties. However, SWCNTs are still very expensive to produce and remain a challenge for the development of more affordable synthesis techniques.

Multiwalled carbon nanotubes (MWCNTs): MWCNTs consist of multiple cylinders and can be considered as a collection of concentric SWCNTs (consisting of multiple layers of graphite rolled in on themselves to form a tube shape) with different diameters. They have closed ends with hemispheres of fullerene molecules. MWCNTs have an outer diameter of 2–100 nm and an inner diameter of 1–3 nm, with a length of 1 to several hundred μm. Intermediate CNTs are DWCNTs composed of two concentric graphene tubes. Perfect CNTs have all carbons bonded in a hexagonal lattice except at their ends. The length and diameter of these structures differ significantly from those of SWCNTs, and, of course, their properties are also very different. The interlayer distance in MWCNTs is close to the distance between graphene layers in graphite, approximately 3.3 Å.

The special case of MWCNTs (double-walled carbon nanotubes, DWCNTs) must be emphasized here because they combine very similar morphology and properties compared to SWCNTs. CNTs can be multiwalled with a central tubule of nanometric diameter surrounded by several graphitic layers separated by approximately 3.4–3.6 Å. DWCNTs were synthesized recently by different methods. DWCNTs are interesting 1D nano-objects with a structure and properties between SWCNTs, MWCNTs, and DWCNTs. Theoretical calculations and high-resolution transmission electron microscopy measurements demonstrated that the interlayer distance between the inner and outer tubes of the DWCNTs can vary from 0.33 to 0.41 nm, which is larger than that of MWCNTs (0.34 nm). Due to the large and variable interlayer distances, DWCNTs demonstrate unique electronic, optical, and mechanical properties. For example, theoretical calculations show that the energy gap of semiconductive DWCNTs can be tuned by changing the interwall distance. A multiwalled carbon nanotube is composed of many nanotubes of different diameters; thus, their properties are not well-defined. A nanorope is composed of many adjacent nanotubes.

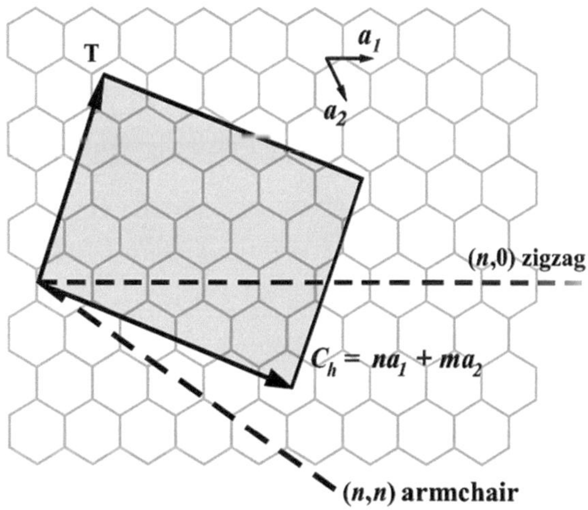

Fig. 5.4 Depicts the rolling of a hexagonal sheet of graphite to form a CNT with a vector structure classification defined by two integers (n, m). The chiral vector for rolling up (Ch = $na_1 + ma_2$), often referred to as the chiral index (n, m), uniquely defines the diameter (d) and the chiral angle (θ) of the SWNT [8, 9]

Different Ways of Wrapping Result in Different Kinds of CNTs

The hexagonal lattice structures of SWCNTs are classified as armchair, zigzag, and chiral. Multiwalled carbon nanotubes (MWCNTs) can be considered as a collection of concentric SWCNTs, consisting of multiple layers of graphite rolled in on themselves to form a tube shape, with different diameters. The structure and rolling of SWCNTs and MWCNTs depend on how the graphene sheet is oriented during rolling. This orientation can be defined by a vector known as the chiral vector, which determines how the graphene sheet is rolled up (Fig. 5.4).

The equations for this are as follows:

$$d = \left[\left(n^2 + nm + m^2\right)^{1/2} \times 0.142\,\text{nm}\right] / \left[3\left(n^2 + m^2\right)^{1/2}\right]$$
$$\theta = \arctan[(3^{1/2} \times m)/(2n + m)],$$

where a^{cc} (0.142 nm) is the nearest neighbor C–C distance.

The chiral index also directly determines the electronic structure of the nanotube: It is metallic when $(n - m)/3$ is an integer (and "semi-metallic" if $n \neq m$), and otherwise, it is semiconducting.

Carbon nanotubes are solely described by the pair of integers (n, m), which is related to the chiral vector. The primitive lattice vectors of the hexagonal lattice are represented by a and b, with r being the roll-up vector and θ representing the chiral angle. As illustrated in Fig. 5.5a–c, three types of CNTs can be characterized by these values: (a) a metallic conducting (10,10) tube referred to as "armchair," (b) a semiconducting (12,7) tube known as "chiral," and (c) a conducting (15,0) tube called "zigzag."

In Fig. 5.5c, a SWCNT with an intramolecular junction connecting two domains is shown: one with metallic properties and the other with semiconductive properties.

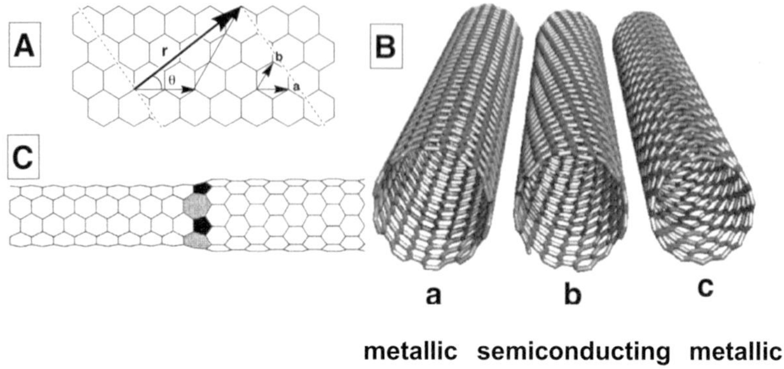

Fig. 5.5 **A** Scheme illustrating the folding procedure for creating nanotube cylinders from planar graphene sheets: here, *a* and *b* represent the primitive lattice vectors of the hexagonal lattice, while *r* represents the roll-up vector, and θ denotes the chiral angle. **B** Idealized representation of defect-free SWCNTs (n, m) with open ends: (a) a metallic conducting (10,10) tube referred to as "armchair," (b) a semiconducting (12,7) tube known as "chiral," and (c) a conducting (15,0) tube referred to as "zigzag." **C** SWCNT with an intramolecular junction connecting two domains: one domain exhibits metallic properties, while the other possesses semiconductive properties

The structures of armchair and zigzag tubes exhibit a high degree of symmetry, as indicated by the arrangement of hexagons around their circumference.

The naming scheme for (n, m) nanotubes can be envisioned as a vector (Ch) within an infinite graphene sheet, describing how to "roll up" the graphene sheet to create the nanotube. *T* represents the tube axis, while a_1 and a_2 denote the unit vectors of graphene in real space. The structure of a SWCNT can be defined using a roll-up vector *r* and a chiral angle θ, with the roll-up vector being expressed as a linear combination of the base vectors *a* and *b* of the fundamental hexagon (where m and n are integers):

$$\mathbf{r} = n\mathbf{a} + m\mathbf{b}$$

Techniques for Producing Carbon Nanotubes: There are several widely used techniques developed for the fabrication and synthesis of CNTs. These techniques are primarily based on gas-phase processes, with the most used methods including:

1. Electric arc discharge
2. Laser ablation
3. Chemical vapor deposition (CVD)
4. High-pressure carbon monoxide disproportionation.

The first two methods involve the vaporization of carbon in an inert atmosphere, while CVD and HiPco rely on the catalytic decomposition of hydrocarbons. Laser ablation and arc discharge techniques were initially used to synthesize CNTs at high temperatures. However, these methods have now been largely replaced by low-temperature CVD methods ($\leq 800\,°C$) that offer greater control over nanotube length, diameter, alignment, density, and orientation.

Advancements in CNT synthesis have enabled the engineering of patterns composed of aligned MWNTs or SWNTs. Such patterned growth is of significant interest in tissue engineering. CNT-based scaffolds offer advantages over the biodegradable synthetic polymers currently used in tissue engineering, as they possess structural integrity and high mechanical stability, supporting developing tissue and withstanding in vivo forces.

Arc Discharge Technique: This commonly used technique involves higher temperatures ($> 1700\,°C$) for CNT synthesis, typically resulting in CNTs with fewer structural defects and expanded structures. In this method, arc discharge occurs between high-purity graphite electrodes, usually water-cooled and with diameters ranging from 6 to 12 mm, separated by 1–2 mm in a chamber filled with helium (500 torr) at sub-atmospheric pressure. (Note: Helium can be replaced by a hydrogen or methane atmosphere.) The process involves the evaporation of carbon molecules in a chamber containing a graphite cathode and anode, along with a certain amount of metal catalyst particles, such as cobalt, nickel, and/or iron.

The CNT synthesis initiates by passing direct current through the chamber, creating an arcing process, and applying approximately 4000 K heat under high pressure. During the arcing process, the anode is consumed, and approximately half of the evaporated carbon solidifies on the cathode (negative electrode) tip, forming a cylindrical, hard deposit or cigar-like structure at a rate of 1 mm/min. The remaining carbon is deposited on the inner core of the chamber, creating cathode soot and chamber soot, which have dark and soft characteristics. These deposits yield either single-walled or multiwalled carbon nanotubes along with nested polyhedral graphene particles. Scanning electron microscopy (SEM) can be employed to differentiate their textures and morphologies. The inner core exhibits bundle-like structures, while the outer gray shell consists of solid graphene layers.

In the arc discharge deposition and synthesis of CNTs, catalyst precursors or non-catalyst precursors can be utilized. For instance, multiwalled carbon nanotubes (MWCNTs) are generally synthesized without the use of catalysts. In contrast, single-walled carbon nanotubes (SWCNTs) utilize different catalyst precursors, which can be a composition of graphite and a metal such as Gd, Co, Ni, Fe, Ag, Pt, Pd, etc., or mixtures of Co, Ni, and Fe with other elements like Co–Pt, Co–Ru, Ni–Y, Fe–Ni, Co–Ni, Co–Cu, Ni–Cu, Fe–No, Ni–Ti, Ni–Y, etc. High yields ($> 90\%$) of SWNTs with an average diameter of 1.4 nm have been obtained when Ni–Y-graphite mixtures were used.

The main advantage of the arc discharge technique lies in its ability and potential for the production of a large quantity of nanotubes. However, one notable disadvantage of this method is the relatively limited control over the alignment (chirality) of the created nanotubes, which is crucial for their characterization and specific applications.

Laser Ablation: In this technique, high-power laser vaporization is employed within a quartz tube containing a block of pure graphite subjected to heating inside a furnace at approximately 1200 °C, all in an argon (Ar) atmosphere. The objective is to vaporize the graphite within the quartz, generating single-walled carbon nanotubes (SWNTs) and introducing metal particles as catalysts to the graphite targets. Research has indicated that the diameter of the nanotubes depends on the applied laser power, with increased laser power resulting in thinner tubes. Conversely, ultrafast (sub-picosecond) laser pulses can produce substantial quantities of SWNTs (approximately 1.5 g/h).

The properties of CNTs synthesized using the laser ablation method, such as the structural and chemical composition of the target material, laser properties (peak power, continuous wave versus pulsed, energy influence, oscillation wavelength, and repetition rate), flow and pressure of the buffer gas, chamber pressure and chemical composition, distance between the target and the substrates, and ambient temperature, can all have an impact. Advantages of this technique include the production of high-purity, high-quality SWNTs, relatively high yields, and low metallic impurities. On the other hand, the main disadvantages are the potential for non-uniform nanotubes with branching, the requirement for high-power lasers, the cost, and relatively low yields.

Chemical Vapor Deposition (CVD): Chemical vapor deposition, or CVD, is one of the standard methods for producing carbon nanotubes. Catalytic chemical vapor deposition (CCVD) is currently the standard technique for synthesizing carbon nanotubes, allowing for CNT expansion and involving the chemical breakdown of a hydrocarbon on a substrate. The process entails a growth mechanism that excites carbon atoms in contact with metallic catalyst particles. Initially, tubes are drilled into silicon and implanted with iron nanoparticles at the bottom. Subsequently, a hydrocarbon such as acetylene is heated and decomposed onto the substrate. The carbon in contact with the metal particles forms nanotubes that mimic the tunnel's shape, resulting in the growth of well-aligned, long CNTs.

All CVD processes are performed at approximately 700 °C, employing common metal catalyst particles such as nickel, cobalt, iron, or combinations thereof. The metal nanoparticles, along with a catalyst support like MgO or Al_2O_3, serve to increase the surface area for enhanced by-product production during the catalytic reaction of pure carbon with the metal particles. The reactor is fueled by two types of gases (with the fluidized bed reactor being the most widely used): a carbon-containing gas (e.g., ethylene, acetylene, methane, or ethanol) and a process gas (e.g., nitrogen, hydrogen, or ammonia) for the CNT growth process. While various CVD processes, such as thermal or plasma-enhanced (PE) oxygen-assisted CVD, water-assisted CVD, microwave plasma (MPECVD), radiofrequency CVD (RF-CVD), or hot-filament (HFCVD), have been studied, CCVD is considered the most used method (Fig. 5.6).

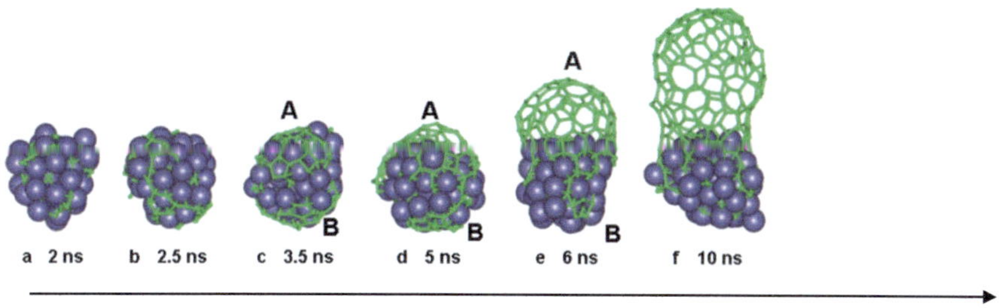

Fig. 5.6 Time-evolution snapshots of carbon nanotube (green) encapsulation and elongation around a metal nanoparticle (purple) over a 10 ns molecular dynamics simulation. Initially, the carbon atoms begin to wrap around the metal cluster (**a–c**), forming partially enclosed structures with domains labeled A (upper cap) and B (lower base). As time progresses (**d–f**), the carbon network extends upward, forming a well-defined nanotube cap and tubular wall, while the metal cluster remains partially exposed at the base. This process illustrates the dynamic formation of carbon nanotubes on metal catalyst surfaces [10]

Purification of CNTs

Oxidative treatment of CNTs with nitric and sulfuric acid results in their fragmentation into shorter pieces. For practical applications, carbon nanotubes must undergo purification and sorting based on length, diameter, and structure. Separation and purification are particularly crucial for controlling the essential properties of carbon nanotubes, especially their conductivity (Fig. 5.7).

The purity of CNTs can be enhanced through advancements in synthesis and purification techniques. Recently, exceptionally pure, metal-free SWNTs have been successfully synthesized using a combination of gas- or vapor-phase oxidation, wet chemical oxidation, centrifugation, filtration, or chromatography. SWNTs can be aligned by applying an electric field during the CVD growth of the nanotubes, while vertically aligned arrays of MWNTs are routinely produced in standard CVD experiments.

Fig. 5.7 Reaction mechanism of oxidative treatment of CNTs

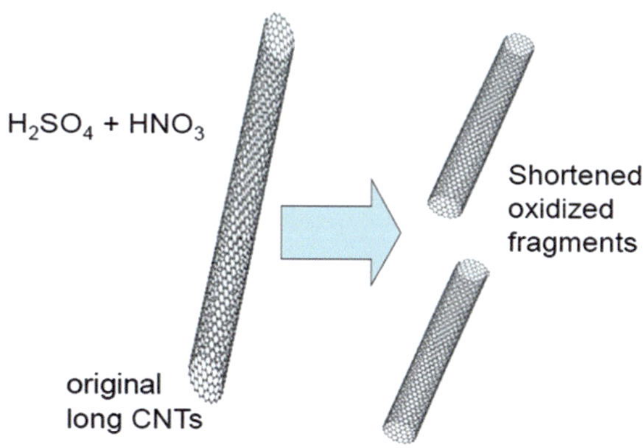

Applications such as biosensors and bioelectronics often require relatively small amounts of materials. However, scenarios may arise where bulk quantities of CNTs are necessary, underscoring the importance of achieving high purity in CNT materials as an essential prerequisite for further progress. In comparison to SWNTs, MWNTs can be obtained in a highly purified form. Typically, SWCNTs contain a significant fraction of carbonaceous impurities and traces of the metal catalyst used in the synthesis process. Consequently, ensuring purity has been a challenge in SWNT manufacturing, mainly due to the absence of widely accepted and established methods for quality assessment regarding commercially available CNT materials.

In biotechnology and biosensors, high-purity materials are expected to play a critical role, even though quality control is currently lacking in the CNT industry. Analytical tools such as solution-phase near-infrared spectroscopy, Raman spectroscopy, thermogravimetric analysis, and combinations of these methods have been employed to evaluate SWNT purity and establish reliable quality assessments for SWNT materials. Efforts to establish industry standards in this field may lead to reforms in current CNT manufacturing practices. For quantitative assessment of the carbonaceous purity of SWNTs, NIR spectroscopy has emerged as the simplest, fastest, most efficient, and unambiguous technique, while thermogravimetric analysis can estimate the metal content.

The Importance of SEM, TEM, and AFM Imaging in Nanomaterials Research

Scanning electron microscopy (SEM) and transmission electron microscopy (TEM) are essential tools for characterizing nanomaterials such as carbon nanotubes (CNTs), due to their ability to reveal detailed structural and morphological information at the nanoscale (Fig. 5.8).

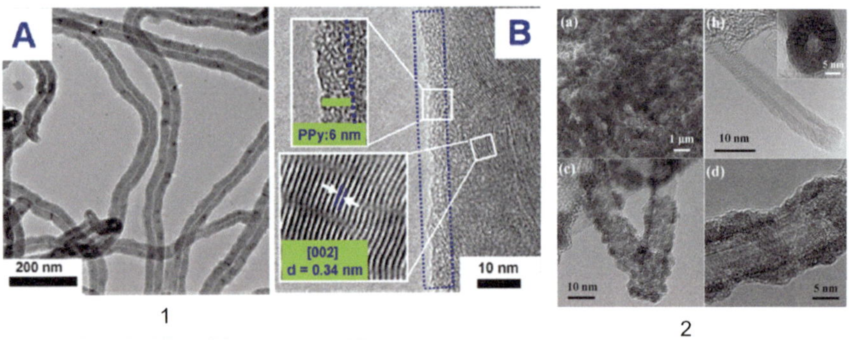

Fig. 5.8 (**1**) High-resolution TEM image of a polypyrrole/multiwall PPy/MWCNT. (**2**) SEM/TEM composite image that beautifully captures buckled and folded multi-walled carbon nanotubes under deformation: the SEM panel illustrates large-scale structures of CNT bundles; the TEM panels, including a cross-sectional inset, reveal individual MWCNT walls and deformation features such as buckling, collapse, and curvature—key indicators of structural flexibility

SEM provides high-resolution surface imaging by scanning a focused electron beam across the sample and detecting secondary or backscattered electrons. This technique excels at visualizing surface topography, bundle formation, entanglement, and macroscopic deformations such as buckling or collapse in CNT networks. SEM is particularly useful for examining large-area distributions and interactions between nanotubes or their integration into devices.

TEM, on the other hand, transmits electrons through an ultra-thin sample, enabling atomic- or near-atomic-level resolution. TEM is indispensable for probing the internal structure of individual CNTs, revealing features such as wall thickness, number of concentric shells (in multiwalled CNTs), defects, kinks, or partial collapses. High-resolution TEM also enables direct observation of lattice fringes, chirality, and local deformation, which are crucial for understanding mechanical and electronic behavior.

Together, SEM and TEM offer complementary insights: SEM highlights surface and network-level organization, while TEM reveals fine structural and crystalline details. These imaging techniques are critical for correlating the physical geometry of CNTs with their functional properties, including electrical conductivity, electron emission, and mechanical flexibility—enabling informed design and optimization of CNT-based nanodevices.

The resolution of atomic force microscopy (AFM) depends on the sharpness of the tip, and it can be significantly improved by using a tip made of a single-wall carbon nanotube (SWCNT). SWCNTs are attached to an AFM probe, allowing for precise measurements and imaging.

Carbon nanotubes can also be utilized as tips in AFM applications. Such a tip can be employed for various purposes, including creating nanoscale patterns through nanolithography or etching materials from a surface in the fabrication of semiconductor chips. In this context, the tip acts like a "nanotweezer," enabling the removal of individual atoms from the surface.

Video clips demonstrate the real-time dynamics of interactions between carbon nanotube tips and silicon or diamond surfaces. The nanotube tip can selectively extract several silicon atoms from a silicon surface. However, when interacting with diamond, the harder diamond surface can damage or destroy the nanotube.

Carbon nanotubes are recognized as some of the stiffest, strongest, and toughest fibers known to date. Stiffness is measured by Young's modulus, which quantifies the rate of stress change concerning applied strain. Young's modulus of the best nanotubes can reach as high as 1000 GPa, approximately five times greater than that of steel. Additionally, nanotubes exhibit a tensile strength of up to 63 GPa, which is around 50 times higher than steel. These exceptional properties, combined with the lightweight nature of carbon nanotubes, make them highly promising for applications in aerospace and other industries.

Biomolecule Interaction with Carbon Nanotubes

The interaction between DNA and carbon nanotubes (CNTs) can be highly specific and depends on the structure of both DNA and CNTs. CNTs can form a parallel array, which can be inserted into the DNA groove. This property can be harnessed for sorting CNTs based on their specific interactions with DNA molecules.

The wrapping of carbon nanotubes by single-stranded DNA has been found to be sequence-dependent and can vary among different types of CNTs. In solution, unsorted CNTs typically appear black (far left), while conducting nanotubes exhibit a pinkish color, and semiconducting ones appear greenish.

Chemical Modification

The conceptual formation of biomolecule-carbon nanotube conjugates and their assembly to create functional devices represents a crucial aspect of future bio nanotechnology. CNT-biomolecule hybrid systems hold significant potential in advancing this field.

Typical defects in a single-walled carbon nanotube (SWCNT) include (a) five- or seven-membered rings in the carbon framework, which deviate from the normal six-membered ring and result in a bent tube. (b) Sp3-hybridized defects (R = H and OH). (c) Carbon framework damage caused by oxidative conditions, leading to holes lined with –COOH groups. (d) Open ends of SWCNTs, terminated with –COOH groups.

In addition to the carboxy termini shown, other terminal groups such as $–NO_2$, OH, H, and =O are also possible. Functionalization possibilities for single-walled carbon nanotubes (SWCNTs) include (a) defect-group functionalization; (b) covalent sidewall functionalization; (c) non-covalent hexahedral functionalization with surfactants; (d) non-covalent exohedral functionalization with polymers; (e) endohedral functionalization, for example, with C60 as shown in Fig. 5.9.

Carbon nanotubes can be filled with molecules within their internal channels. CNTs can be functionalized with various biomolecules, organic materials, and various nanoparticles without the need for covalent coupling. Open-ended CNTs

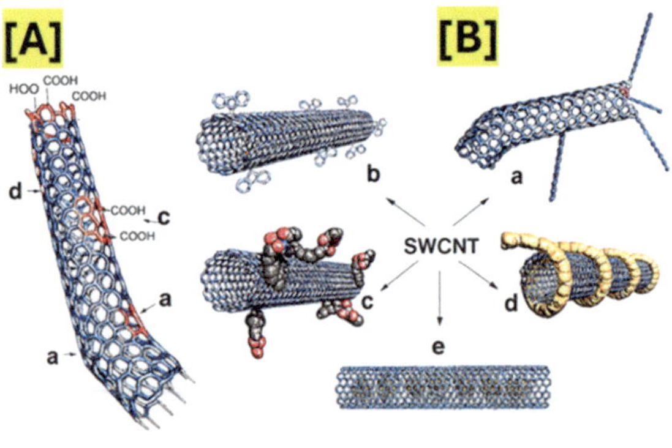

Fig. 5.9 **A** Schematic representation of a single-walled carbon nanotube (SWCNT) functionalized with carboxylic acid groups (–COOH) at different positions along the nanotube wall and tips, labeled as sites a–d. **B** Illustrations of various chemical functionalization and structural modification strategies for SWCNTs: (a) junction formation, (b) covalent sidewall functionalization, (c) non-covalent surface adsorption, (d) polymer wrapping, and (e) encapsulation within the nanotube channel

provide internal cavities (1–2 nm in diameter) capable of accommodating organic molecules and biomolecules of respective sizes.

For example, research has shown that open SWCNTs can host fullerenes and small proteins like cytochrome c within their internal cavities, resulting in the stable immobilization of proteins in bioactive conformations. DNA can also enter the cavities of CNTs, and the transport of DNA through a single MWCNT channel has been directly observed using fluorescence microscopy (Fig. 5.10).

Molecular dynamics simulations have suggested that DNA can be encapsulated inside SWCNTs in a water solute environment through an extremely rapid dynamic interaction process, provided that the tube size exceeds a certain critical value. The snapshots in the simulation illustrate the process of an oligonucleotide (8 adenine bases) interacting with a (10,10) carbon nanotube at 0, 30, 100, and 500 ps.

Carbon nanotubes (CNTs) can be wrapped by polymers or biopolymers, resulting in water-soluble hybrid materials. For example, amylose chains coil around single-walled carbon nanotubes, allowing them to dissolve in water. Polymeric shells around CNTs provide dispersion and solubility. In TEM and SEM images, non-modified

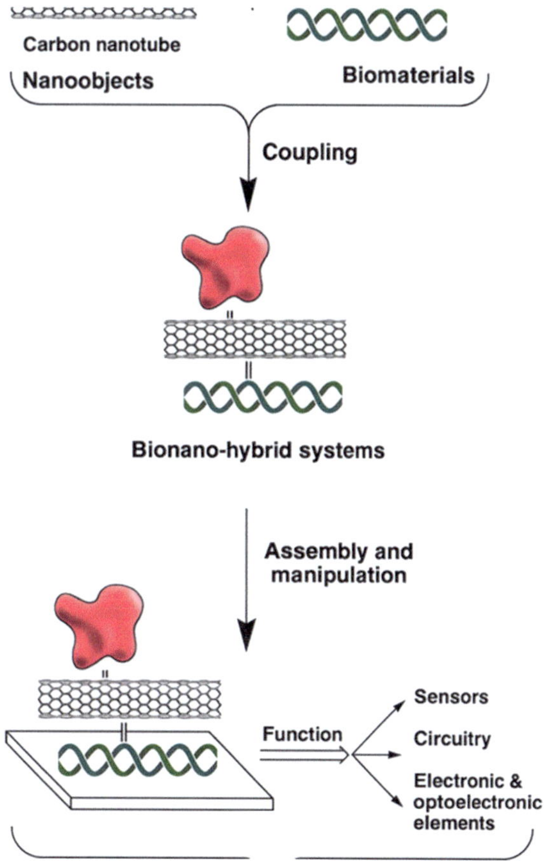

Fig. 5.10 Functionalization possibilities for single-walled carbon nanotubes (SWCNTs)

SWCNTs show a high degree of bundling, making them insoluble in water and prone to precipitation. In contrast, polymer-coated SWCNTs form aqueous dispersions.

π-Conjugated pyrene units serve as "sticky" anchor groups, facilitating the binding of molecules to the external side of CNT walls. These units enable the surfactant to bind to the graphitic sidewalls of CNTs. Positively charged ammonium groups further contribute to the water-soluble dispersion of functionalized CNTs.

CNTs can also undergo covalent functionalization with organic and bio-organic molecules. Oxidation of CNTs generates carboxylic groups at the ends and on the walls, which can be used for electrostatic binding of positively charged polymers or covalent coupling of organic and bio-organic molecules.

Various techniques, such as layer-by-layer electrostatic self-assembly, diimide-activated amidation, and carbodiimide coupling reactions, enable the binding of proteins, nanoparticles, quantum dots, and DNA to CNTs. These methods provide control over the functionalization of CNTs, making them versatile platforms for various applications.

Despite these advancements, site-specific and non-symmetric modification of CNTs remains a challenging goal, with different groups bound to the walls and ends, as well as the opposite ends of CNTs. Site-specific immobilization of Au nanoparticles (10 nm) on MWCNTs can be achieved through various methods (Fig. 5.11).

Specific attachment of Au nanoparticles to the ends (a) or only one end of CNTs (b) can be achieved (Fig. 5.12). Covalent modification of CNT sidewalls results in the formation of defect sites with sp3-electron hybridization, significantly altering the electron energy near the Fermi level.

The electron energy disturbance extends over a large distance (more than 1 nm) despite the structural deformation being limited to the vicinity of the defect site. Consequently, even low concentrations of defect sites introduced into the sidewalls by covalent modification can significantly decrease the conductivity of CNTs. This provides an effective means to control the conductivity of CNTs through covalent modifications. Long-range periodicity in the covalent functionalization of sidewalls of SWCNTs is possible. Nucleophilic addition of 1,8-diazabicycloundec-7-ene (Bingel reaction) effectively occurs at domains of the sidewalls separated by long distances (several nanometers), resulting in a pattern on the SWCNTs. The covalently bound substituents generate sp3-electron hybridization sites that locally decrease the conductivity of the sidewalls, allowing the visualization of the generated pattern by STM.

This effect is attributed to the long-range electronic effects generated in the conductive sidewalls upon the formation of sp3-electron defects. The long-range periodic alterations of the properties of the CNT sidewalls were further exploited to generate polymer single crystals periodically grown on CNTs. Disk-shaped polymeric single crystals of polyethylene or nylon-6,6 were periodically generated on the sidewalls of CNTs separated by approximately 20–70 nm distances, which can be controlled by the deposition procedure. Since the polymers can be easily removed from the sidewalls by etching-dissolution, the polymer can operate as a

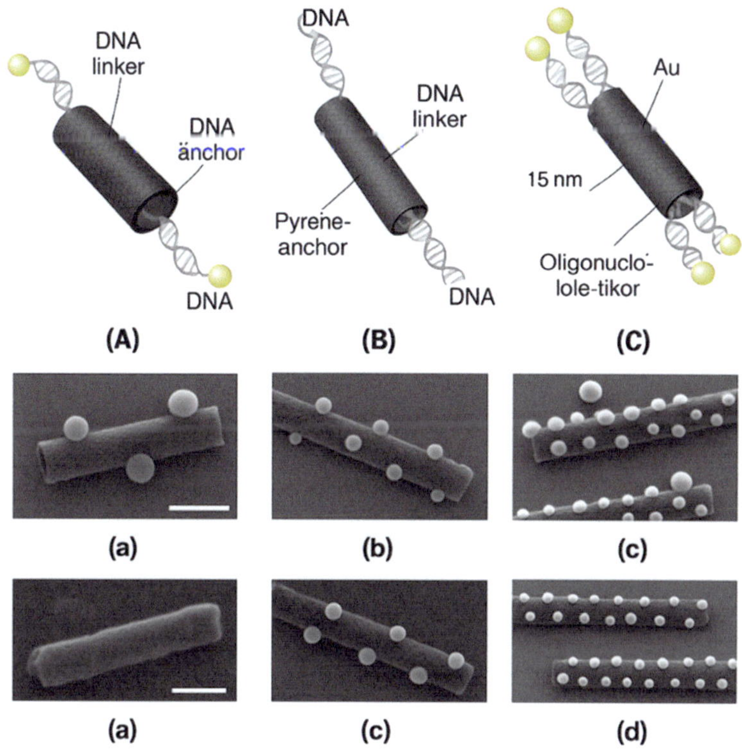

Fig. 5.11 **A** Use of DNA linkers covalently attached to the end-carboxylic groups. **B** DNA linkers bound to the sidewalls via pyrene-anchor groups. Corresponding images (a) and (b) show SEM pictures of the resulting MWCNTs/Au nanoparticles assemblies, respectively. **C** Site-specific parallel immobilization of Au nanoparticles of different sizes (30 and 15 nm) using two oligonucleotide linkers associated with the ends and sidewalls of the MWCNTs, respectively. SEM images (c) and (d) depict the resulting MWCNTs/Au nanoparticles assembly

mask, providing a new route for the periodic modification of CNTs with different functionalities.

Electronic and Bioelectronic Applications

Ordered deposition of metal nanoparticles on solid supports and their use as catalytic centers for the growth of CNTs resulted in the formation of aligned arrays of CNTs with controlled diameters and lengths. An SEM image of the Fe/Mo-cluster dots deposited on a surface by microcontact printing before the growth of the SWCNTs. Another SEM image shows the SWCNTs array grown from the patterned catalyst dots.

Covalent attachment of carbon nanotubes via carbodiimide coupling led to densely packed arrays of SWCNTs on hydrophilic strips, with the alignment of the SWCNTs controlled by the direction of rinsing and drying. AFM images of SWCNTs

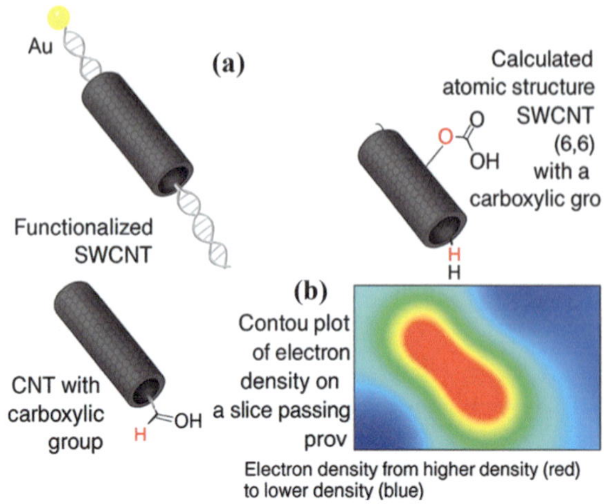

Fig. 5.12 Calculated distribution of electron density in a SWCNT with a carboxylic group covalently bound to the sidewall. **a** Shows the calculated atomic structure of a SWCNT (6,6) with a carboxylic group. **b** Displays a contour plot of the electron density on a slice passing through the carboxylic group. Red, yellow, green, and blue colors on the contour plot indicate electron density from higher density to lower density

(Fig. 5.13) adsorbed on hydrophilic strips of amino-functionalized siloxane are generated by a microcontact printing technique (a and b show the images at different magnifications).

Carbon Nanotubes as Electron Conductors in Bioelectrocatalytic Systems and Enzyme-Based Electrochemical Sensors: Many advantages of CNTs-modified electrodes stem from their highly specific area. Numerous publications discuss the electrocatalytic activity of CNTs-modified electrodes for various electrochemical reactions. A significant portion of the electrocatalytic activity observed in CNTs-modified electrodes originates from defect sites, particularly edge-plane-like defect sites. However, some reports attributing "special" electrocatalytic activity to CNTs-modified electrodes are incorrect, as the catalytic activity often arises from metal impurities remaining in the CNTs. Figure 5.14a illustrates the electrical contacting

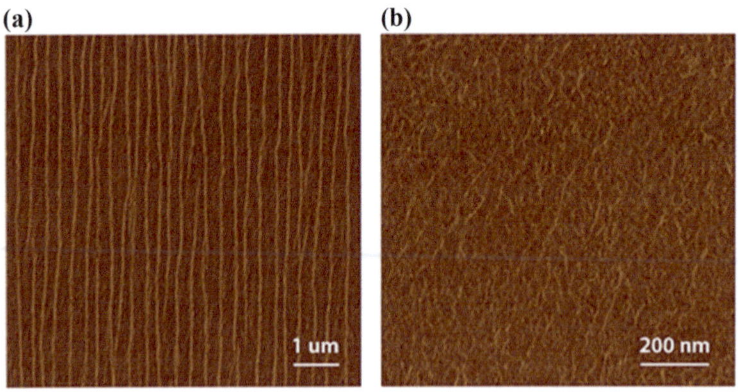

Fig. 5.13 AFM images of SWCNTs

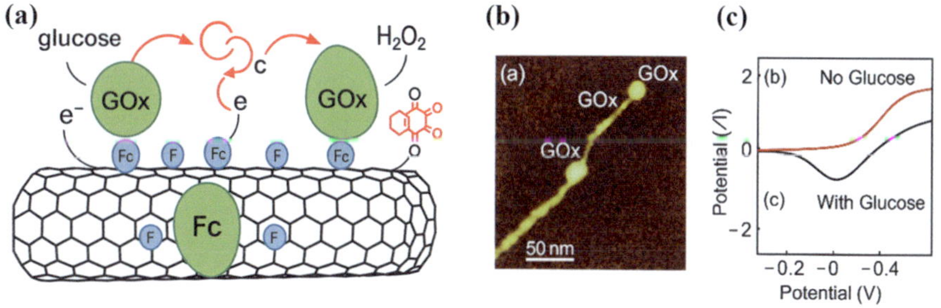

Fig. 5.14 **a** Electrical contacting of GOx loaded onto SWCNT sidewalls through a diffusional mediator (ferrocene monocarboxylic acid). **b** AFM image of a SWCNT loaded with GOx on the sidewall. **c** Voltammetric responses of GOx-modified SWCNTs

of glucose oxidase (GOx) loaded onto SWCNT sidewalls via a diffusional mediator, ferrocene monocarboxylic acid. An AFM image showing a SWCNT with GOx immobilized on its sidewall is presented in panel b. Voltammetric responses of GOx-modified SWCNTs using ferrocene monocarboxylic acid as a diffusional electron mediator are shown in panel c: (b) in the absence of glucose and (c) in its presence.

Carbon Nanotubes as Supports for Immunosensing and DNA-Sensing Arrays

Figure 5.15 illustrates an ultrasensitive multiplex electronic biosensor based on a carbon nanotube nanoelectrode array. The insets on the right show applications in DNA and antigen detection. The system enables amplified electrochemical DNA analysis using oligonucleotide-functionalized MWCNTs that bind poly-guanine oligonucleotides acting as redox-active labels.

Electrochemically controlled and addressable formation of Pd nanoparticles on MWCNTs arrays is shown in Fig. 5.16. Covalent coupling of SWCNTs to a silicon surface through highly conjugated molecular bridges allowed the formation of electrical junctions between the metallic or semiconductive CNTs and the semiconductive support, yielding assemblies with unique electronic properties. The various biological molecules conjugated to carbon nanotubes are also explored.

5.3 Conducting Polymers

Conducting polymers (CPs) are defined as organic polymers capable of conducting electricity. They can exhibit metallic conductivity or behave as semiconductors. Historically, polyaniline (PANI) was the first conducting polymer described in the mid-nineteenth century by Henry Letheby, who investigated the electrochemical behavior of aniline in acidic media. He observed the color change between the reduced form (colorless) and the oxidized form (dark blue).

In the 1950s, many researchers began to investigate the electrochemical behavior and charge-transfer properties of numerous polycyclic aromatic compounds [11].

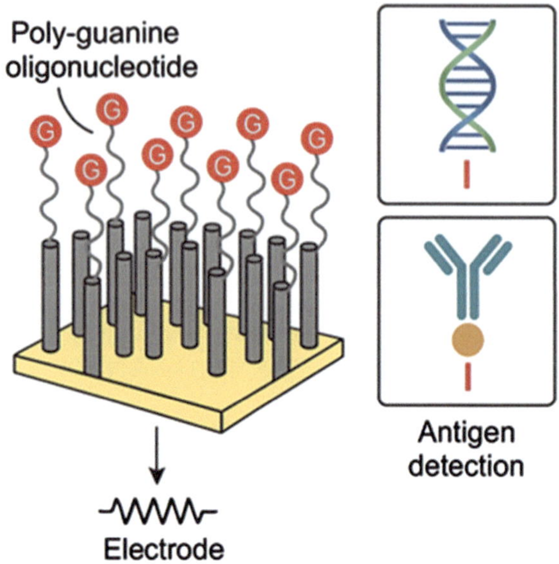

Fig. 5.15 Schematic representation of an ultrasensitive multiplex electronic biosensor based on a carbon nanotube nanoelectrode array. Vertically aligned multiwalled carbon nanotubes (MWCNTs) are functionalized with poly-guanine oligonucleotides, which serve as redox-active labels for amplified electrochemical DNA detection. Insets on the right illustrate applications in DNA and antigen detection

In 1954, researchers at Bell Labs and elsewhere reported organic charge-transfer complexes with resistivities as low as 8 Ω/cm [12, 13]. In the early 1970s, researchers demonstrated that salts of tetrathiafulvalene exhibited almost metallic conductivity, while superconductivity was demonstrated in 1980 [14].

In 1977, Alan J. Heeger, Alan MacDiarmid, and Hideki Shirakawa [15] reported similar high conductivity in oxidized iodine-doped polyacetylene. For this groundbreaking research, they were awarded the 2000 Nobel Prize in Chemistry "for the discovery and development of conductive polymers."

Research into the charge-transfer properties of many semiconducting, conducting, and superconducting polymers is ongoing, even after almost two centuries, especially regarding their wide applications in the development of various devices, from solar cells and light-emitting diodes (LEDs) to biosensors based on different transduction mechanisms (e.g., optical, electrochemical, or electronic).

CPs can be classified as shown in Table 5.1.

Conductive polymers can be prepared in different ways. Most conductive polymers are prepared by oxidative coupling of monocyclic precursors. Such reactions entail dehydrogenation:

$$n\,\mathrm{H} - [\mathrm{X}] - \mathrm{H} \rightarrow \mathrm{H} - [\mathrm{X}]_n - \mathrm{H} + 2(n-1)\,\mathrm{H}^+ + 2(n-1)\,\mathrm{e}^-.$$

There are two main methods used to synthesize conductive polymers: chemical synthesis and electro(co)polymerization. The former involves connecting carbon–carbon bonds of monomers by subjecting simple monomers to various conditions, such as heating, pressing, exposure to light, and catalysts. This method allows for a high yield but results in a lower purity of the final product.

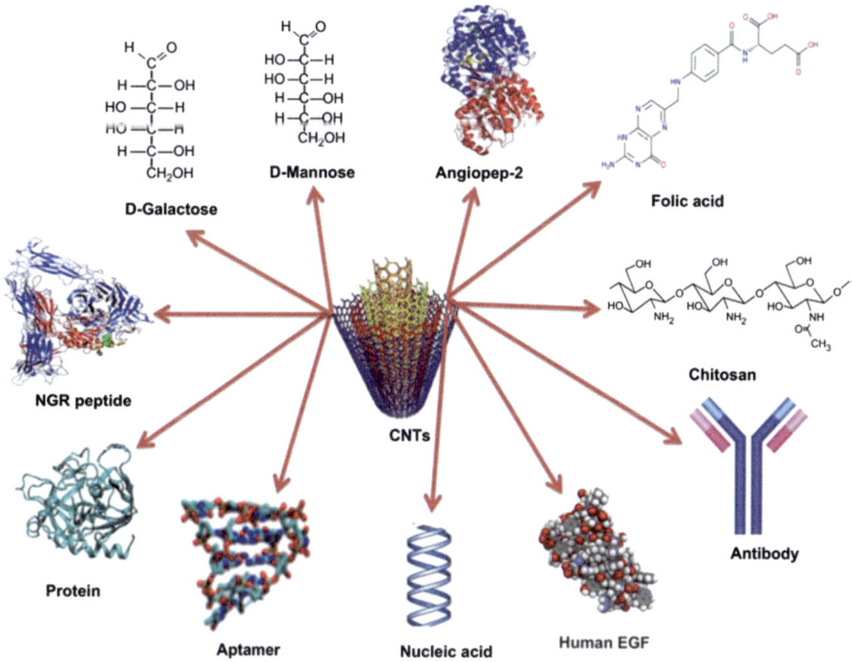

Fig. 5.16 Schematic illustration of various biomolecules and ligands that can be functionalized onto carbon nanotubes (CNTs) for biosensing and biomedical applications. The central CNT structure is shown radiating connections to diverse molecules, including carbohydrates (D-galactose, D-mannose), peptides (NGR peptide, Angiopep-2), proteins, aptamers, nucleic acids, antibodies, folic acid, chitosan, and human epidermal growth factor (EGF), highlighting the versatility of CNT surface modification for targeted detection and therapeutic strategies

Table 5.1 CPs to Conductive Polymers based on main chain content, aromatic cycles, double bonds, and the presence of heteroatoms

Main chain content	No heteroatom	Heteroatom present	
		Nitrogen-based	Sulfur-based
Aromatic cycles	Poly(fluorene)s Polyphenylenes Polypyrenes Polyazulenes Polynaphtalenes	Poly(pyrrole)s (PPY) Polycarbazoles Polyindoles Polyazepines Polyanilines (PANI)	Poly(thiophene)s (PT) Poly(3,4-ethylenedioxythiophene) (PEDOT) Poly(p-phenylene sulfide) (PPS)
Double bonds	Poly(acetylene)s (PAC)		
Aromatic cycles and double bonds	Poly(p-phenylenevinylene) (PPV)		

Conversely, electro(co)polymerization takes place in an electrochemical cell where a solution containing reactants or monomers is placed. The redox reaction is promoted by sweeping or stepping a certain voltage (or voltage range). Both potentiostatic and potentiodynamic methods in this class are at the forefront of conductive polymer synthesis.

Among conductive polymers (CPs), polyaniline (PANI) is certainly the most widely used, serving as both a conducting polymer and a semiconductor. PANI can exist in three idealized oxidation states:

- **Leucoemeraldine** (fully reduced)—white/clear and colorless $(C_6H_4NH)_n$.
- **Emeraldine**—green for the emeraldine salt, blue for the emeraldine base $([C_6H_4NH]_2[C_6H_4N]_2)_n$.
- **(Per)nigraniline** (fully oxidized)—blue/violet $(C_6H_4N)_n$.

The form of polyaniline often referred to as emeraldine base (EB) is neutral. When doped (protonated), it is called emeraldine salt (ES), with the imine nitrogen protonated by an acid. Protonation helps delocalize the otherwise trapped diiminoquinone-diaminobenzene state. Emeraldine base is considered the most useful form of polyaniline due to its high stability at room temperature. Upon doping with an acid, the resulting emeraldine salt form of polyaniline becomes highly electrically conductive. Leucoemeraldine and pernigraniline are poor conductors, even when doped with an acid.

PANI has found applications in the construction of various types of biosensors, including those involving antigenic, genomic, and enzymatic biorecognition interactions. For instance, Liu and colleagues used three-dimensional macroporous conducting polymer PANI to fabricate an electrochemical immunosensor for quantifying AFP, a cancer liver biomarker, as shown in Fig. 5.17. The electroactive surface area of the 3D macroporous PANI-modified electrode is 3.4 times larger than that of the planar PANI-modified electrode. This larger surface area allows for the immobilization of numerous AFP antibodies. In this sensor, the electrochemical signal is generated by the inherent redox signal from PANI, without the need for additional redox reagents. Under optimum conditions, the porous PANI-based immunosensor exhibited a wide linear range from 0.01 to 1000 pg mL^{-1} and a detection limit of 3.7 fg mL^{-1} [16].

Saberi and co-workers reported a nanogenosensor capable of detecting microRNA-16, which, when overexpressed, can be considered a cancer biomarker, at a detection limit of 0.1 nM. In this example, the authors modified the electrode surface with PANI, gold nanoparticles to dramatically enhance both the electroactive area and its conductivity, and alkaline phosphatase as a label for analytical detection [17].

Chen et al. also designed a sandwich-type DNA biosensor for sensing the specific IS6110 DNA fragment of Mycobacterium tuberculosis (MTB), which has been confirmed as a key molecular marker for the diagnosis of TB [18]. The capture probe was immobilized onto the surface of the Au electrode via amino-Au affinity. Subsequently, the target DNA was added to the Au electrode, and the PANI-rGO/AuNPs composite was used as a redox detector probe to generate the voltammetric signal

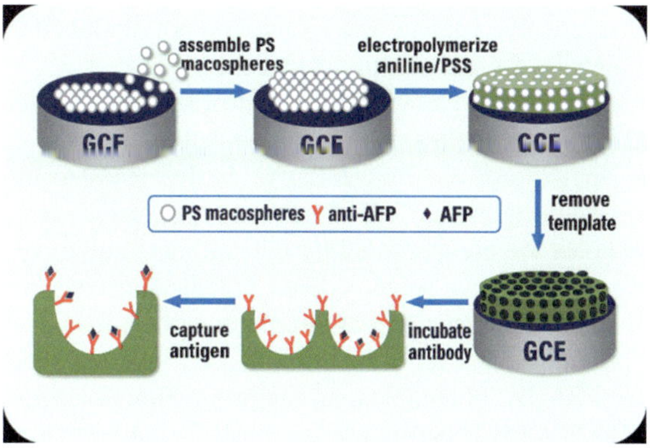

Fig. 5.17 Schematic representation of a molecularly imprinted electrochemical biosensor fabrication process on a glassy carbon electrode (GCE). Polystyrene (PS) macrospheres are first assembled on the GCE surface, followed by electropolymerization of aniline/PSS to form a porous conductive matrix. After template removal, anti-AFP (α-fetoprotein) antibodies are incubated to form specific binding sites. The resulting sensor captures AFP antigens through selective recognition. Symbols indicate PS macrospheres (○), anti-AFP (Y), and AFP (♦)

due to appropriate electrochemical activity. The authors believed that the PANI/rGO/AuNP nanohybrid enhanced the electron transfer rate and increased the sensitivity of the biosensor. In fact, incorporating PANI with rGO provided many advantages, including increased immobilization of PANI, which improves the electrochemical signals and creates a good platform for other modifications. This is probably because of the numerous amino groups of PANI. Moreover, the high conductivity and good adsorption properties of AuNPs improved the electrical characteristics and tightly immobilized detector probes via a strong Au–S bond. This genosensor presented a linear range of 0.1 pM–10 nM and a limit of detection of 50 fM.

In enzymatic amperometric detection, such as for glucose as a diabetes biomarker, Tamer et al. used a PANI/gold nanorod (GNR) composite to fabricate an electrochemical enzymatic nano-biosensor for glucose recognition. Glucose oxidase (GOx) enzyme was immobilized on the modified PANI/GNR electrode with covalent bonding via glutaraldehyde. This composite provided an appropriate matrix for enzyme immobilization due to its large specific surface area and higher electroactivity. Their results showed a calibration curve with a linear range from 17.6 μM to 1 mM, a sensitivity of 13.8 $\mu A\ mM^{-1}\ cm^{-2}$, and a limit of detection (LOD) of 5.8 μM [19].

Another conducting polymer (CP) widely used in the construction of electronic biosensors is poly(3,4-ethylenedioxythiophene) polystyrene sulfonate (PEDOT:PSS). PEDOT:PSS is a polymer mixture of two ionomers. One component in this mixture is composed of sodium polystyrene sulfonate, which is a sulfonated polystyrene. Part of the sulfonyl groups are deprotonated and carry a negative charge. The other component, poly(3,4-ethylenedioxythiophene) (PEDOT), is a conjugated polymer and carries positive charges and is based on polythiophene. Together, these charged macromolecules form macromolecular salt.

A typical material for organic electrochemical transistors (OECTs) is the conducting polymer PEDOT:PSS. The semiconducting PEDOT is p-type doped (oxidized in the language of electrochemistry), which leads to mobile holes that can hop from one chain to another, resulting in a hole current once a drain voltage is applied. These holes are compensated by the sulfonate anions of PSS, which, in terms of solid-state physics, can be considered as ionized acceptors.

In sensing applications considering affinity interactions, Kim et al. demonstrated an OECT-based immunosensor for the detection of the prostate-specific antigen/α1-antichymotrypsin (PSA-ACT) complex, an important biomarker for prostate hyperplasia Fig. 5.18. The PEDOT:PSS film was successively modified by the AuNPs-PSA pAb, PSA-ACT complex, PSA mAb, and ProLinkerTM molecules Fig. 5.18. The use of AuNPs-PSA pAb enabled amplified the electrochemical signals between the PSA-ACT complex/PSA mAb and the OECT device, leading to a low detection limit of 1 pg/mL (well below the cut-off limit of 4 ng/mL) [20].

Furthermore, He et al. successfully grafted anti-*Escherichia coli* O157:H7 antibodies onto the PEDOT:PSS channel via silane treatment and demonstrated that OECTs can be used to detect bacteria (*E. coli* O157:H7). The detection limit of the bacteria sensor was 10^{-3} cfu mL^{-1}. Additionally, after functionalizing PEDOT:PSS films with glucose oxidase enzyme, OECTs showed a linear response to glucose ranging from 1.1 to 16.5 mM, covering the human body blood glucose level range of 3.85–8.25 mM [21].

5.4 Plasmonic Nanomaterials and Photonic Crystals

Plasmonic nanomaterials are defined as nanometer-sized materials whose electron density can couple with electromagnetic radiation of wavelengths much larger than the nanomaterial itself, considering the nature of the dielectric-metal interface between the medium and the surface. What sets these nanomaterials apart from normal surface plasmons is that plasmonic nanomaterials also exhibit interesting scattering, absorbance, and coupling properties based on their geometries and relative positions. These unique properties have made them a focus of research in many applications, including solar cells, spectroscopy, signal enhancement for imaging, cancer treatment, and bio(sensor) development. For plasmon and plasmonic resonance, you can recall the concept from the metal nanoparticles section. However, plasmons consist of oscillations of free electrons as a consequence of the formation of a dipole in the material due to electromagnetic waves. Indeed, the electrons move back to the material to restore its initial state, while the light waves oscillate, leading to a constant shift in the dipole that forces the electrons to oscillate at the same frequency as the light.

Plasmonic nanomaterials (P-NMs) are typically classified into two categories: noble metal-based MNPs and MNP-carbon nanomaterials, including MNP-CNT and MNP-GRP. Both types of P-NMs exhibit excellent plasmonic and optical properties. Among these, noble metals, such as Au, Ag, and Pt-based MNPs, are well-established and widely applied in various fields due to their stable plasmonic properties and particle stability. Additionally, the plasmonic properties of MNPs can be easily tuned and controlled through particle structure or the distance between each MNP. On the other hand, the tunable infrared (IR) plasmonic property of GRP has also garnered interest due to its excellent optical properties, such as enhancing light-matter interactions and integrating mid-IR photonics.

Oh et al. tuned the plasmonic properties of AgNPs by controlling their particle size. In this case, the size was altered using the seed-mediated growth method, and as the Ag ion concentration increased, larger AgNPs were produced. Furthermore, the plasmonic absorbance of AgNPs was redshifted, resulting in a change in the solution color from yellow to red. Choi et al. tuned the plasmonic properties of AuNPs by controlling the assembly of AuNPs. Typically, spherical AuNPs have a plasmonic absorbance at approximately 525 nm. However, in this study, the authors synthesized chains of AuNPs using N-(3-Dimethylaminopropyl)-N′-ethylcarbodiimide (EDC), and as the EDC concentration increased, the AuNP chains elongated, altering their structure. Consequently, the absorbance behavior of the AuNP chain changed and shifted to longer wavelengths (redshifted). This phenomenon is attributed to the plasmonic coupling effect between closely contacting AuNPs, which induced a secondary plasmonic band observed at approximately 625 nm.

P-MNs can be employed for both localized surface plasmon resonance (LSPR) and plasmonic resonance energy transfer (PRET)-based biosensors. The underlying principle of PRET nanospectroscopy is the energy transfer between the plasmon resonance of a metal nanoparticle and the molecular resonant peaks (i.e., absorption peaks) of the conjugated biomolecule. This interaction leads to resonant quenching dips in the Rayleigh scattering spectrum of the particle, as depicted in Fig. 5.18 [22].

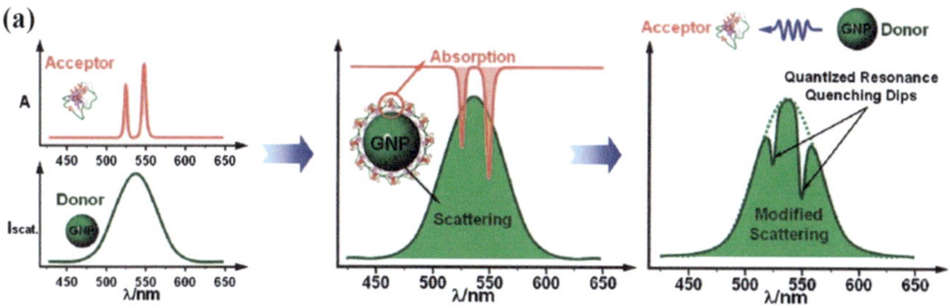

Fig. 5.18 Schematic illustration of localized surface plasmon resonance (LSPR) and plasmonic resonance energy transfer (PRET)-based biosensors utilizing gold nanoparticles (GNPs) as donors. (Left) Independent spectral profiles of the acceptor molecules (absorption peaks, red) and the GNP donor (scattering spectrum, green). (Center) When acceptor molecules are brought into close proximity with the GNP surface, their absorption bands overlap with the donor's LSPR scattering profile, initiating PRET and resulting in resonance quenching. (Right) The modified GNP scattering spectrum shows quantized resonance quenching dips at the acceptor's absorption wavelengths, enabling sensitive and specific molecular detection through spectral fingerprinting

For instance, Lee et al. detected an infectious TB antigen using TB antibody-modified Au-GRP and quantum dots (QD) through a PRET-based fluoro-immunoassay. In this study, a sandwich structure was formed with Au-GRP and QD due to the presence of the TB antigen (i.e., CFP-10), and the fluorescence enhancement of QD was measured in response to the antigen concentration. The limit of detection (LOD) was estimated to be 4.5 pg/mL [23].

Photonic crystals are defined as optical nanostructures that periodically change their refractive index, affecting the propagation of light. This is similar to how the structure of natural crystals gives rise to X-ray diffraction and how the atomic lattices (crystal structure) of semiconductors influence their electron conductivity. Photonic crystals can be fabricated in 1D, 2D, and 3D forms. One-dimensional photonic crystals can be made by depositing thin film layers on top of each other. Two-dimensional photonic crystals can be created through photolithography or by drilling holes in a suitable substrate. Fabrication methods for three-dimensional photonic crystals include drilling at various angles, stacking multiple 2D layers, direct laser writing, or inducing self-assembly of spheres in a matrix and then dissolving the spheres.

Photonic crystals have found extensive use in the development of waveguide-based biosensors. For instance, Buswell and colleagues demonstrated specific detection of streptavidin to biotin using a photonic crystal waveguide biosensor. The binding of streptavidin to immobilized biotin caused a red shift of 0.86 nm in the cut-off wavelength for a 2.5 nm streptavidin film. The bulk sensitivity of the presented waveguide biosensor was 88 nm/RIU (Refractive Index Unit). To enhance sensitivity, the authors explored a modification to the waveguide sensor. Instead of removing one row of holes, they reduced the size of one line of holes to a fraction of their original size. They used a lattice period (Λ) of 446 nm with a hole radius of 0.37Λ for the exterior holes and 0.22Λ for the holes in the waveguide line. With this device design, the cut-off wavelength remained unaltered compared to the previous design, but the modes were shifted to higher frequencies within the bandgap. This modification resulted in an improved sensitivity of up to 120 nm/RIU, representing a 40% enhancement compared to the sensor with a regular waveguide [24].

Most of the biosensors based on photonic crystals have primarily focused on proving the concept, while performing real-time multiplexing analysis for further integration into lab-on-chip devices for self-monitoring or remote medicine remains a challenging task.

5.5 Quantum Dots

Quantum dots consist of semiconductor nanoparticles exhibiting both optical and electronic properties that usually differ from bulk materials due to quantum mechanics. While irradiated by UV light, an electron in the quantum dot can be promoted to a higher energy level (excitation). Indeed, the excitation corresponds to the transition from the valence to the conducting band for a certain semiconducting material. In the case of a semiconducting quantum dot, this process corresponds to the

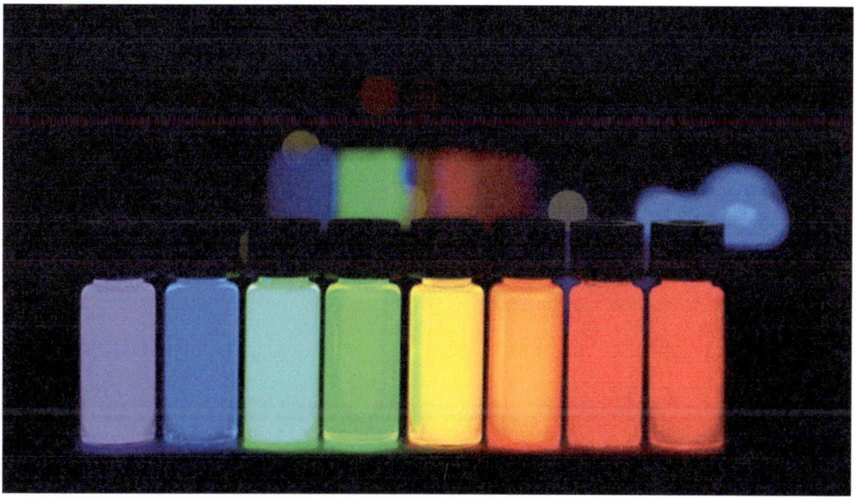

Fig. 5.19 Photograph of quantum dot solutions exhibiting bright photoluminescence under UV light. The image shows a series of vials containing quantum dots that emit distinct colors ranging from violet to red, corresponding to their size-dependent emission wavelengths. This tunable fluorescence property is widely exploited in biosensing, imaging, and optoelectronic applications

transition of an electron from the valence band to the conductance band. The excited electron can drop back into the valence band releasing its energy by light emission. This light emission (photoluminescence) usually appears as depicted in Fig. 5.19. The color of that light depends on the energy difference between the conductance band and the valence band, or transition between discretized energy states when band structure is no longer a good definition in QDs.

Quantum dots exhibit intermediate properties between bulk semiconductors and discrete atoms/molecules. Their optoelectronic properties change as a function of both size and shape. They can potentially be applied in the development of solar cells, single-electron transistor, quantum computing, cell biology, bio(sensors) based on optical, electronic and electrochemical transduction.

About their optical properties, QDs usually adsorb energy ($E = h\nu$) under UV-light irradiation, thus resulting in the promotion (excitation) of an electron from the valence to the conduction band (forming a hole in the valence band). When the excited electron returns to its ground state, it can emit the *excitation energy* (emitted photon) as light, which is defined as fluorescence, as shown below. Simplifying, the energy of the emitted photon can be understood as the sum of the band gap energy between the valence band and the conduction band, the confinement energies of the hole and the excited electron, and the bound energy of the exciton (the electron–hole pair), as described in Fig. 5.20.

It should be highlighted that the confinement energy is size-dependent; hence, it can be tuned during the synthesis, offsetting the energy level for both absorption and fluorescence. Larger dots absorb and emit lower energy compared to smaller dots that

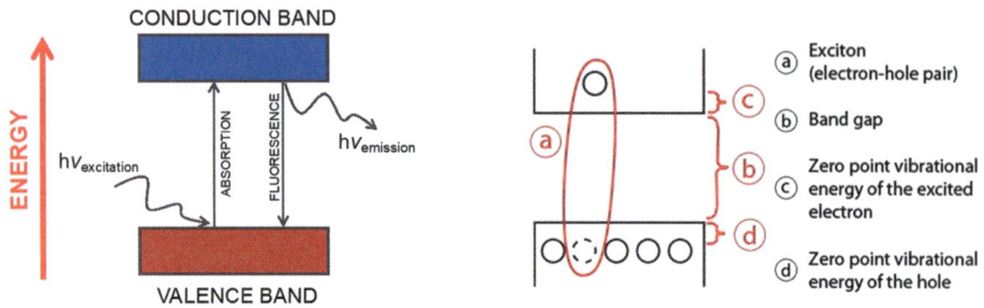

Fig. 5.20 Energy band diagrams illustrating excitonic processes in semiconductors. (Left) A photon excites an electron from the valence band to the conduction band, generating an exciton (electron–hole pair). The exciton may recombine radiatively (emitting a photon) or undergo non-radiative Auger recombination, where the recombination energy is transferred to a third carrier instead of being emitted as light. (Right) Detailed diagram showing the exciton state (a), band gap energy (b), zero-point vibrational energy of the excited electron (c), and zero-point vibrational energy of the hole (d). These processes are fundamental to understanding the optical responses of nanomaterials in biosensing applications

emit higher energy. The fluorescence quantum yield can be improved by creating a shell around the quantum dot with a semiconductor having a larger bandgap so that non-radiative surface recombination pathways can be limited, as well as Auger recombination.

QDs can be synthesized using several methods, both chemical and physical. For instance, colloidal synthesis, classified as a wet chemistry process, is carried out by heating the solution at high temperatures, decomposing the precursors that form monomers, proceeding through nucleation and the generation of nanocrystals. Temperature is a key control factor in this synthetic process because the temperature should be high enough to lead to the annealing of atoms and their rearrangement. Another key factor is the concentration of monomers. At higher concentrations, smaller particles grow faster than larger ones, resulting in a more focused size distribution (nearly monodisperse suspension). Over time, the monomer concentration tends to decrease, and the critical size becomes larger than the average size (defocusing). Typical dots are made of binary compounds such as lead sulfide, lead selenide, cadmium selenide, cadmium sulfide, cadmium telluride, indium arsenide, and indium phosphide. Dots may also be made from ternary compounds such as cadmium selenide sulfide. Furthermore, recent advances have been made that allow for the synthesis of colloidal perovskite quantum dots. As reported in the literature, QDs can also be fabricated through self-assembling processes that naturally occur in quantum wells (discovered in the 1970s) due to monolayers' fluctuations in the wells. Another synthetic pathway is the well-known Stranski–Krastanov method. Self-assembled quantum dots nucleate spontaneously under certain conditions during molecular beam epitaxy (MBE) and metalorganic vapor-phase epitaxy (MOVPE) when a material is grown on a substrate to which it is not lattice-matched [25].

Alternatively, QDs can be synthesized using the plasma method as a gas-phase approach, especially adopted in the formation of binary compounds with covalent bonds. For example, silicon (Si) and germanium (Ge) quantum dots have been synthesized using nonthermal plasma. The size, shape, surface, and composition of quantum dots can all be controlled in nonthermal plasma.

In some reports, researchers have evidenced the possibility of genetically engineered bacteriophage viruses serving in the biosynthesis of quantum dots. This has been particularly shown for M13 bacteriophage viruses. In addition, the liquid crystalline structures of wild-type viruses (Fd, M13, and TMV) are adjustable by controlling the solution concentrations, solution ionic strength, and the external magnetic field applied to the solutions. Consequently, the specific recognition properties of the virus can be used to organize inorganic nanocrystals, forming ordered arrays over the length scale defined by liquid crystal formation. Based on this concept, Lee et al. reported the fabrication of self-assembled, highly oriented, self-supporting films from a phage and ZnS precursor solution. This system allowed them to vary both the length of the bacteriophage and the type of inorganic material through genetic modification and selection [26].

QDs are also synthesized using electrochemical methods where a template is electrodeposited onto the electrode surface, leading to QD layers self-assembling in the template, which is finally etched away.

Upon closer examination of the literature, new synthetic methods for QDs preparation are published every day, depicting a nanotechnology field that is continuously evolving, not only on the application side but also from a synthetic point of view.

Quantum dots (QDs) are commonly integrated into the development of electrochemical biosensors, serving either as signal amplifiers or labels, depending mostly on the biorecognition interaction (e.g., enzymatic reactions, immunometric assays, or DNA/DNA, RNA/RNA, and aptamer interactions, etc.). In the case of immunometric reactions, Liu and co-workers proposed an amperometric enzyme immunosensor with amplified sensitivity for the determination of the alpha-fetoprotein (AFP) tumor marker. They achieved this by employing a layer-by-layer assembly of poly(3,4-ethylenedioxythiophene) (PEDOT)/nano-Au/Azure I/ZnSe QDs on the Pt electrode surface [27]. ZnSe QDs were used to improve the loading of anti-AFP antibodies. Finally, horseradish peroxidase (HRP) was employed to block sites against non-specific binding and amplify the current signal of the antigen–antibody reaction. The immobilization of ZnSe QDs on the Azure I layer led to an increase of about 130% in redox peak currents. This signal enhancement could be explained by the fact that ZnSe QDs intercalated into PEDOT/nano-Au/Azure I acted as a conducting wire or an electron-conducting tunnel, facilitating direct electron transfer. The described immunosensor displayed high sensitivity, fast analytical time, a relatively low detection limit of 1.1 fg mL^{-1}, and an especially broad linear response to AFP in a range from 5×10^{-5} to 250 ng mL^{-1}.

More recently, Farschi and co-workers reported a sandwich-type voltammetric immunosensor for the monitoring of prostate-specific antigen (PSA) using citrate-AgNPs decorated GQDs nano-ink modified flexible paper substrate (Fig. 5.21). In sandwich-type immunosensors, the capture antibodies are usually immobilized on a modified electrode, and the biological target (antigen) is sandwiched between the immobilized primary antibodies (Ab1) and secondary/signal antibodies (Ab2). In the described immunosensor, citrate-capped silver nanoparticles modified GQDs (citrate-AgNPs-GQDs) nano-ink were directly written on the cellulose paper surface (ivory sheet and photographic paper) and modified by the cysteamine functionalized AuNPs (CysA-AuNPs). This provided a good and high external area for the immobilization of biotinylated anti-PSA Ab1 in a large amount. To detect the target, PSA antigen was dropped on the sensing area of the prepared immunosensor and incubated for 30 min at 4 °C. Consequently, HRP-anti-PSA Ab2 was immobilized on the sensing area of the engineered sensor. Finally, the electrochemical behavior of the sandwich-type immunoassay was evaluated by DPV technique. The obtained quantification limit for the first immunodevice (ivory sheet/citrate-AgNPs-GQDs nano-ink/CysA-AuNPs/Ab1/BSA/PSA/HRP-Ab2) was 0.07 μg L^{-1}, and the linear range for the calibration plot was from 0.07 to 60 μg L^{-1}. Also, the achieved quantification level for the second immunodevice (photographic paper/citrate-AgNPs-GQDs nano-ink/CysA-AuNPs/Ab1/BSA/PSA/HRP-Ab2) was 0.05 μg L^{-1}, with a linear range of 0.05–10 μg L^{-1} [28].

Kokkinos et al. reported a voltammetric immunosensor for the detection of the CRP biomarker based on graphite screen-printed electrodes (SPEs) modified with bismuth citrate using PbS QDs as the labeling material. This sandwich-type immunoassay involved the physisorption of anti-CRP Ab1 on the electrode surface, sequential immunoreactions with CRP and biotinylated anti-CRP Ab2, and finally, a reaction with streptavidin-conjugated PbS QDs. The CRP determination was performed with the acidic dissolution of the PbS QDs and ASV detection of the released Pb^{2+} ions. The quantification was performed at bismuth nanodomains formed on the sensor surface during the electrolytic preconcentration step, as bismuth citrate was reduced to metallic bismuth simultaneously with the deposition of Pb on

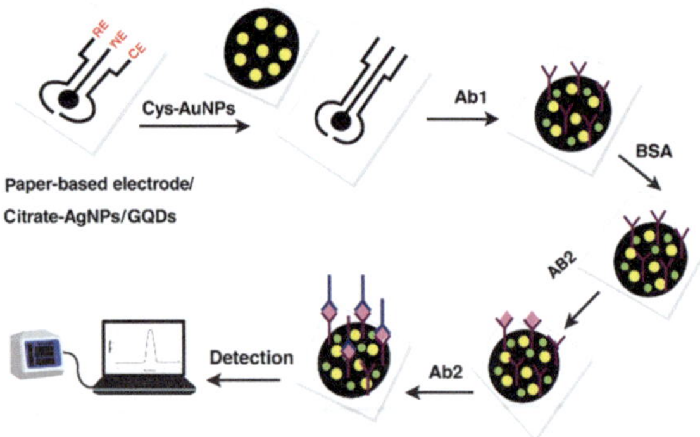

Fig. 5.21 Schematic illustration of a paper-based electrochemical immunosensor for prostate-specific antigen (PSA) detection. The sensor is constructed using a paper-based electrode modified with citrate-stabilized AgNPs/GQDs and cysteine-functionalized AuNPs. Sequential steps include immobilization of the capture antibody (Ab1), blocking with BSA, binding of the target PSA, and addition of a secondary antibody (Ab2) for signal amplification. The resulting immune complex is analyzed via electrochemical detection, enabling sensitive and specific quantification of PSA

the surface of the immunosensor. The biosensor response was linear over the range of 0.2–100 ng mL^{-1} while the detection limit was 0.05 ng mL^{-1} [29].

Since QDs behave similarly to organic dyes or fluorescent proteins, they have also been exploited in bioimaging applications and optical biosensors. For instance, monitoring intracellular molecular motion with QDs, such as EB1 (a microtubule-associated protein) and single myosin V (an actin-associated protein), has been demonstrated with high spatial and temporal resolution. As shown in Fig. 5.22, after conjugation *i*th QDs, the continuous movement of individual EB1 proteins along the microtubule structure was monitored optically, while the mean velocity was also evaluated. This result highlights the ability of QDs to monitor the dynamic behavior of biomolecules over long periods of time and their applications in studying complex biological structures and their functions. In medical applications, imaging is an important technique for disease diagnosis as well as therapy. Many imaging techniques, such as magnetic resonance imaging (MRI), positron emission tomography (PET), computed tomography (CT), and optical imaging, have been developed for clinical applications. Recently, the design and fabrication of imaging probes with multimodalities have become an active research area where different contrast agents can be fused into a single nanoparticle system. QDs have been extensively investigated in this area because of their unique optical properties and the ease of conjugating additional contrast agents to them.

One method for using QDs in biosensing is to create a donor/acceptor complex, which exhibits switching capability via fluorescence resonance energy transfer (FRET). FRET between donor and acceptor molecules has been extensively used

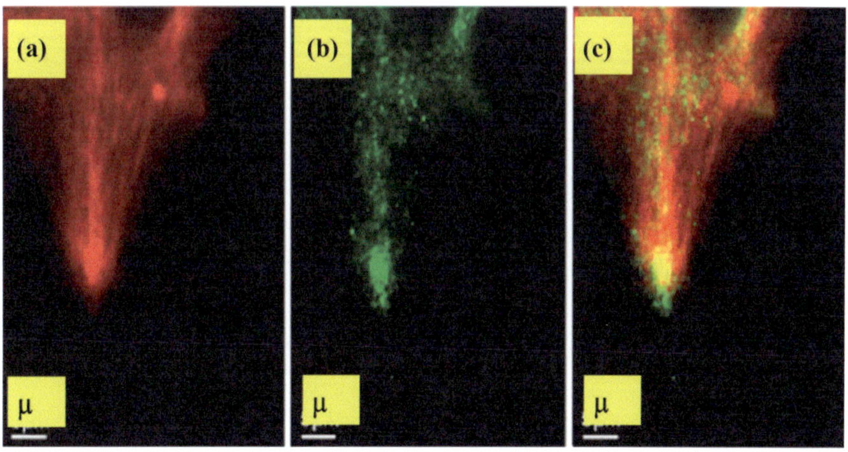

Fig. 5.22 Fluorescence microscopy images showing cellular localization of two different fluorescent markers. **a** Red fluorescence signal, **b** green fluorescence signal, and **c** merged image showing co-localization of red and green signals, indicating overlapping regions in yellow. Scale bars represent micrometers (μm). These images demonstrate dual-labeling for analyzing spatial distribution and interactions within biological samples

in biophysical and biochemical studies to probe ligand-receptor binding and molecular structural changes. However, the main limiting factor in the performance of QDs as FRET donors lies in their size because the FRET efficiency depends on the center-to-center separation between the donor and acceptor. In this regard, several studies have demonstrated the effective use of QD FRET donors to detect small analytes by utilizing a common strategy that relies on conjugating QDs to target binding receptors, which can be either proteins, antibody fragments, or DNA aptamers. The QD conjugates are then exposed to appropriate acceptor-labeled target analogs, which are brought in proximity to the QDs by binding to the receptors. In this initial state, the QD-donor photoluminescence is quenched by efficient FRET to the proximal acceptor-labeled analogs. The presence of the target then displaces bound analogs from the surrounding conjugated receptors, and this is detected through a reduction in FRET efficiency and the concomitant increase in QD photoluminescence (Fig. 5.23).

This strategy has been utilized for detecting diverse analytes. An interesting biosensor showing reversible FRET has been fabricated by connecting CdSe/ZnS core/shell QDs with a photoactivatable species that functions as the reversible FRET acceptor [29].

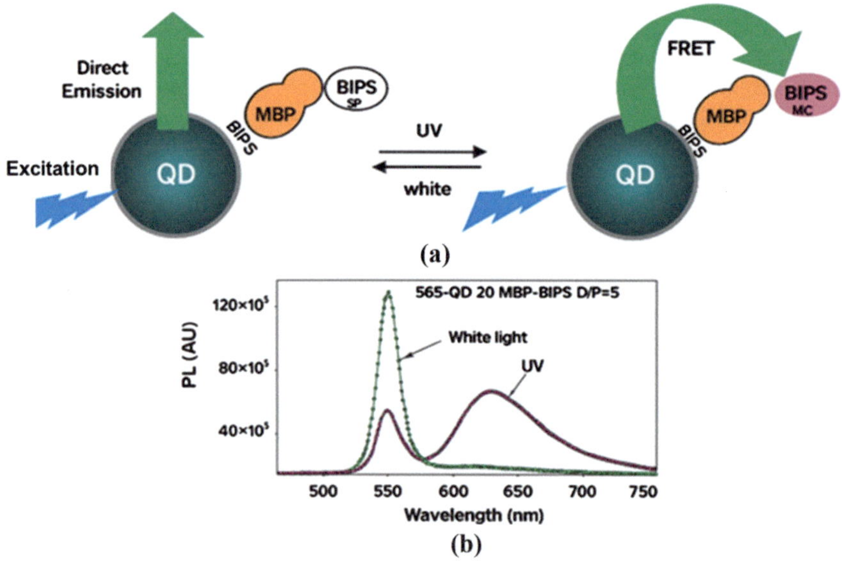

Fig. 5.23 **a** Schematic representation of the light-induced reversible switching between direct quantum dot (QD) emission and Förster Resonance Energy Transfer (FRET) via photochromic protein interactions. Under white light, the BIPS chromophore remains in the SP (spiropyran) state, leading to direct QD emission. Upon UV exposure, BIPS converts to the MC (merocyanine) state, enabling FRET from the QD to BIPS. **b** Photoluminescence (PL) spectra of the QD-MBP-BIPS system under white and UV light, showing modulation of emission intensity and spectral shift due to photoswitchable energy transfer. Reproduced from [29, 30]

QDs are connected to the photochromic molecule 1′,3-dihydro-1′-(2-carboxyethyl)-3,3-dimethyl-6-nitrospiro-[2H-1-benzopyran-2,2′-(2H)-indoline] (BIPS) via a bridge of maltose binding protein (MBP). Exposure to ultraviolet light catalyzes the photoconversion of BIPS from the colorless spiropyran to the colored merocyanine form, which functions as the FRET acceptor and therefore modulates QDs emission. The photoconversion is reversible, with white light converting merocyanine back to the spiropyran form. Quenching of QD emission at 555 nm and enhancement of BIPS emission at 650 nm via FRET occur upon exposure to ultraviolet light [29].

Well-controlled, reversible switching events were demonstrated by alternating the illumination light source between white and ultraviolet light. Incorporation of an emission unit that can be modulated via a biological stimulus enables the creation of photochromically switched devices or sensors, where QD emission modulation presets the device below some predetermined critical threshold. The detection limits of analytical processes based on FRET can be as low as 10 ppt with a linear dynamic range from 0.1 to 1000 ppt. To create multicolor optical coding, several QDs can be conjugated with the biorecognition element as a sort of barcode reader in an ELISA plate.

References

1. Mie, G. Beiträge zur Optik trüber Medien, speziell kolloidaler Metallösungen. *Ann. Phys.*, **330**, 377–445 (1908).
2. F. Antonii. *Panacea Aurea-Auro Potabile.* (Hamburg: Bibliopolio Frobeniano, 1618).
3. Kimling, J., Maier, M., Okenve, B., Kotaidis, V. & H. Ballot, and A. P. Turkevich Method for Gold Nanoparticle Synthesis. *J. Phys. Chem. B* **110**, 15700–15707 (2006).
4. Brust, M., Walker, M., Bethell, D., Schiffrin, D. J. & Whyman, R. Synthesis of thiol-derivatised gold nanoparticles in a two-phase liquid-liquid system. *J. Chem. Soc. Chem. Commun.* 801–802 (1994) https://doi.org/10.1039/C39940000801.
5. González-Ballesteros, N., Rodríguez-González, J. B., & Rodríguez-Argüelles, M. C. Harnessing the wine dregs: An approach towards a more sustainable synthesis of gold and silver nanoparticles. *J. Photochem. Photobiol. B.* **178**, 302–309 (2018).
6. Gan, P. P., Ng, S. H., Huang, Y., Fong, S. & Li, Y. Green synthesis of gold nanoparticles using palm oil mill effluent (POME): A low-cost and eco-friendly viable approach. *Bioresour. Technol.* **113**, 132–135 (2012).
7. Rykov, V. T., Lukyanovich, V. M., & Radushkevich, L. V. The natural sorbents of the Far East: Communication 1. Electron-microscopic investigations of natural sorbents. *Bull. Acad. Sci. USSR, Div. Chem. Sci.* **1**, 395–397. (1952).
8. M. S. Dresselhaus et al. Science of Fullelerenes and Carbon nanotubes. Ac. Press (1996). p.756–760.
9. Ball, Philip. Roll up for the revolution. Nature 414.6860 (2001): 142–145.
10. Ding, F., Rosén, A. & Bolton, K. Molecular dynamics study of the catalyst particle size dependence on carbon nanotube growth. *J. Chem. Phys.* **121**, 2775–2779 (2004).
11. Naarmann, H. *Polymers, Electrically Conducting. In Ullmann's Encyclopedia of Industrial Chemistry, (Ed.).* (2000).
12. Okamoto, Yoshikuko and Brenner, W. Polymers. In *Organic Semiconductors.* 125–158 (1964).
13. Akamatu, Hideo; Inokuchi, Hiroo; Matsunaga, Y. Electrical Conductivity of the Perylene–Bromine Complex. *Nature.* **173**, 168–169 (1954).

14. Ferraris, John S; Cowan, D. O.; Walatka, V.; Perlstein, J. H. Electron transfer in a new highly conducting donor-acceptor complex. *J. Am. Chem. Soc.* **95**, 94 (1973).
15. Audenaert, M., Gusman, G. & Deltour, R. Electrical conductivity of I2-doped polyacetylene. *Phys. Rev. B* **24**, 7380–7382 (1981).
16. Liu, S., Ma, Y., Cui, M. & Luo, X. Enhanced electrochemical biosensing of alpha-fetoprotein based on three-dimensional macroporous conducting polymer polyaniline. *Sensors Actuators B Chem.* **255**, 2568–2574 (2018).
17. Saberi, R. S., Shahrokhian, S. & Marrazza, G. Amplified electrochemical DNA sensor based on polyaniline film and gold nanoparticles. *Electroanalysis* **25**, 1373–1380 (2013).
18. Chen, Y. *et al.* A polyaniline-reduced graphene oxide nanocomposite as a redox nanoprobe in a voltammetric DNA biosensor for Mycobacterium tuberculosis. *Microchim Acta* **184**, 1801–1808 (2017).
19. Tamer, U., Seçkin, A. İ., Temur, E. & Torul, H. Fabrication of Biosensor Based on Polyaniline/Gold Nanorod Composite. *Int. J. Electrochem.* **2011**, 1–7 (2011).
20. Kim, D.-J. *et al.* Organic electrochemical transistor based immunosensor for prostate specific antigen (PSA) detection using gold nanoparticles for signal amplification. *Biosens. Bioelectron.* **25**, 2477–2482 (2010).
21. Liao J, Si H, Zhang X, L. S. Functional Sensing Interfaces of PEDOT:PSS Organic Electrochemical Transistors for Chemical and Biological Sensors: A Mini Review. *Sensors (Basel)* **19**, 218 (2019).
22. Choi, Y., Kang, T. & Lee P, L. Plasmon Resonance Energy Transfer (PRET)-based Molecular Imaging of Cytochrome c in Living Cells. *Nano Lett.* **9**, 85–90 (2009).
23. Lee, J., Takemura, K. & Park, E. Y. Plasmonic nanomaterial-based optical biosensing platforms for virus detection. *Sensors (Switzerland)* **17**, 1–12 (2017).
24. Buswell, S. C., Wright, V. A., Buriak, J. M., Van, V. & Evoy, S. Specific detection of proteins using photonic crystal waveguides. *Opt. Express* **16**, 15949–15957 (2008).
25. Hulicius, E., Hospodková, A. and Zíková, M. *Quantum Dots. In Metalorganic Vapor Phase Epitaxy (MOVPE). (eds S. Irvine and P. Capper)* (2019).
26. Lee, S.-W., Mao, C., Flynn, C. E., & Belcher, A. M. (2002). Ordering of Quantum Dots Using Genetically Engineered Viruses. *Science (80-).* **296**, 892–895 (2002).
27. Liu, K., Zhang, J., Liu, Q., & Huang, H. Electrochemical immunosensor for alpha-fetoprotein determination based on ZnSe quantum dots/Azure I/gold nanoparticles/poly(3,4-ethylenedioxythiophene) modified Pt electrode. *Electrochim. Acta* **114**, 448–454. (2013).
28. Farschi, F., Saadati, A. & Hasanzadeh, M. A novel immunosensor for the monitoring of PSA using binding of biotinylated antibody to the prostate specific antigen based on nano-ink modified flexible paper substrate: efficient method for diagnosis of cancer using biosensing technology. *Heliyon* **6**, e04327 (2020).
29. Kokkinos, C., Prodromidis, M., Economou, A., Petrou, P. & Kakabakos, S. Disposable integrated bismuth citrate-modified screen-printed immunosensor for ultrasensitive quantum dot-based electrochemical assay of C-reactive protein in human serum. *Anal. Chim. Acta* **886**, 29–36 (2015).
30. Medintz, I. L., Trammell, S. A., Mattoussi, H., & Mauro, J. M. (2004). Reversible modulation of quantum dot photoluminescence using a protein- bound photochromic fluorescence resonance energy transfer acceptor. J. Am. Chem. Soc. **126**(1), 30–31. https://doi.org/10.1021/ja037970h

Chapter 6
Biofunctionalization and Surface Chemistry

Biofunctionalization refers to a specialized form of surface modification involving the immobilization of biomolecules—such as proteins, peptides, polysaccharides, or bioactive drugs—onto solid supports. A variety of immobilization strategies exist, and the selected method significantly influences the biological activity and stability of the resulting functionalized material. Effective biofunctionalization of surfaces and nanomaterials requires a thorough understanding of the chemical reactions and interfacial phenomena occurring between the solid support and the modification medium.

Surface science holds particular importance in the fields of heterogeneous biocatalysis and other electrochemical, optical, and piezoelectric biosensors. In this chapter, we will provide a summary of the main surface modification strategies.

6.1 Covalent and Non-covalent Coupling Strategies

Covalent coupling relies on the formation of a single covalent bond between the polymer surface and the biomolecule. This technique can be used to immobilize small biomolecules, as well as medium and large molecules that possess the appropriate functional group(s). Since the coupling occurs via specific functional groups, the orientation of biomolecules can also be controlled by the user. The main covalent coupling strategies are outlined below.

Carbodiimide Coupling: This coupling strategy is widely employed for immobilizing proteins, DNA, or aptamers on surfaces. Carbodiimides facilitate the binding between the carboxyl groups of a support and the amino function of a protein. Commonly used carbodiimides include the water-soluble 1-ethyl-3-(3-dimethylaminopropyl)carbodiimide (EDC) for aqueous cross-linking and the water-insoluble dicyclohexylcarbodiimide (DCC) for non-aqueous organic synthesis

A. Simonian et al., *Biosensors as Analytical Tools for the 21st Century*,
https://doi.org/10.1007/978-3-032-01726-0_6

methods. N-hydroxysuccinimide (NHS) can be used in conjunction with carbodiimide to enhance immobilization efficiency.

EDC reacts with carboxylic acid groups to create an active O-acylisourea intermediate that can be readily displaced by nucleophilic attack from primary amino groups in the reaction mixture. The primary amine forms an amide bond with the original carboxyl group, releasing a soluble urea derivative as an EDC by-product. The O-acylisourea intermediate is unstable in aqueous solutions; if it fails to react with an amine, it undergoes hydrolysis, regenerating the carboxyls, and releasing an N-unsubstituted urea. EDC cross-linking is most efficient under acidic conditions (pH 4.5) and should be conducted in buffers free of extraneous carboxyls and amines. MES buffer (4-morpholinoethanesulfonic acid) is a suitable carbodiimide reaction buffer. While phosphate buffers and neutral pH (up to 7.2) conditions are compatible with the reaction chemistry, they exhibit lower efficiency. Increasing the amount of EDC in the reaction solution can compensate for this reduced efficiency. N-hydroxysuccinimide (NHS) or its water-soluble counterpart (Sulfo-NHS) is often included in EDC coupling protocols to enhance efficiency or create stable (amine-reactive) intermediates. EDC couples NHS to carboxyls, forming an NHS ester that is considerably more stable than the O-acylisourea intermediate, allowing for efficient conjugation to primary amines at physiologic pH.

Glutaraldehyde Coupling: This procedure facilitates the covalent bonding of two amino groups, one on the surface and the other within the protein, through the formation of an imine bond. The reaction proceeds as primary amino groups react, leading to the formation of intermediate Schiff's bases, which are subsequently converted into stable and complex cross-links.

Affinity Coupling: This strategy enables the creation of (bio)affinity bonds between an activated support (e.g., with lectin, avidin, metal chelates) and a specific group (a tag) within the protein sequence (e.g., carbohydrate residue, biotin, histidine). This method allows for the control of biomolecule orientation to prevent enzyme deactivation and/or active site blocking. Several affinity methods have been described for immobilizing enzymes, including (strept)avidin–biotin, lectin-carbohydrate, and metal cation-chelator interactions. In particular, (strept)avidin–biotin interactions occur between modified proteins (typically biotinylated through the esterification of lysine residues or bioengineering protocols) immobilized on surfaces modified with (strept)avidin, a protein with four interaction sites, although usually only a maximum of two can be used due to steric hindrance. The strong affinity between a metal cation and a chelator like nitrilotriacetic acid (NTA), iminodiacetic acid (IDA), or poly(histidine) tag can also be employed to develop enzymatic biosensors. This method is based on the principles of Immobilized Metal Affinity Chromatography (IMAC), currently used for protein purification and separation. Immobilizing the metal ion on a chromatographic resin through chelation allows for the separation of histidine-tagged proteins from untagged proteins.

Non-covalent immobilization strategies rely on weaker interactions like hydrogen bonding, Van der Waals forces, π-π stacking, and more. Although they do not offer a highly reliable method for biosensor construction, they are widely exploited, especially considering their scalability for industrial production. Proteins can also be entrapped in a polysaccharide-based gel (e.g., alginate, chitosan, or agarose). In contrast to synthetic polymers like polyacrylamide, these matrices are biocompatible, nontoxic, provide a natural microenvironment for the enzyme, and allow for sufficient electron accessibility to shuttle between the enzyme and the electrode.

Another method, widely used in the 80s for biosensor production, involves the inclusion of proteins in a carbon paste made with nujol and pressed within a hollow electrode. Additionally, using inorganic host matrices such as silica or clays serves as an alternative to organic polymers commonly employed for immobilizing enzymes in biosensors. Cationic and anionic clays are two-dimensional layered inorganic solids with open structures that facilitate interactions with enzymes and the intercalation of redox mediators. These clays make suitable host matrices for enzyme immobilization in biosensors, forming biohybrids.

6.2 Specifics of Nanomaterials Biofunctionalization

Nanoscale materials, with their unique properties ranging from 1 to 200 nm, provide an excellent foundation for electronic and optical signal transduction, as well as the development of advanced bioelectronic and biosensing devices. However, the application of nanoparticles (NPs) in analytical chemistry faces challenges related to biocompatibility and biological recognition abilities. To overcome these limitations, biofunctionalization of nanomaterials plays a crucial role in enhancing their biocompatibility for biomolecule immobilization, tissue and cell interactions, and improving specificity for biological recognition. These biofunctional NPs have paved the way for stable biosensing systems with remarkable selectivity and reproducibility. Moreover, they exhibit a synergistic effect, combining catalytic activity, conductivity, and biocompatibility, enabling accelerated signal transduction and achieving rapid, highly sensitive responses through signal amplification. The increasing demand for ultrasensitive bioassays and the trend toward miniaturization have fueled significant interest in the biofunctionalization of nanomaterials. Consequently, researchers are actively exploring suitable methods to functionalize nanomaterials with biomolecules, including proteins, DNA, small organic molecules, polymer films, and even living cells.

Two main approaches have been employed for the functionalization of NPs. The first approach involves non-covalent interactions, such as physical adsorption and biomolecule entrapment around the NPs. These methods utilize electrostatic interactions, π-π stacking, or van der Waals forces to efficiently immobilize biomolecules without disrupting the conjugated skeleton or compromising the electronic properties of the NPs. The second approach is based on covalent interactions, which involve linking biomolecules to the functional groups present on the surface of NPs. Covalent functionalization methods can be broadly categorized into three strategies: direct chemical reactions, linker strategies, and click chemistry. Covalent binding is generally preferred over non-specific physisorption due to its superior stability and reproducibility in surface functionalization processes.

Functional nanomaterials offer diverse opportunities for advancing the field of miniaturization, introducing novel properties and functions, and enabling complex technical developments. The conjugation of nanoparticles (NPs) with biomolecules holds tremendous potential for achieving exceptional signal transduction in the development of electronic or optical biosensors. Optical detection holds several advantages in biosensor design, including high sensitivity, a wide dynamic range, and multiplexing capabilities. NP probes, such as fluorescence energy-transfer nanobeads and quantum dots (QDs), surpass organic dyes and fluorescent proteins in terms of signal brightness, photostability, and the emission of multicolored light. On the other hand, electrochemical assays based on molecular nanoprobes are appealing due to their low cost, high sensitivity, and simplicity. For instance, an electrochemical DNA assay can detect amplified electrochemical signals from microbeads embedded with electroactive molecules, achieving a sensitivity of approximately 100 aM. Furthermore, NPs, being in direct contact with the environment, can function as chemical and biological sensors for the single-molecule detection of biomolecules in electrochemical assays.

This paragraph primarily focuses on the fundamental strategies employed in the biofunctionalization of nanomaterials with biomolecules. Additionally, it highlights some noteworthy applications of biofunctionalized nanomaterials as exceptional electronic or optical signal transducers in the realm of bioanalysis.

Physical Adsorption: The biofunctionalization of nanoparticles (NPs) with biomolecules often involves the straightforward process of biomolecule adsorption. This method encompasses a wide range of biomolecules, from small organic substances to large protein or enzyme molecules. In the case of NPs stabilized by anionic ligands like carboxylic acid derivatives, citrate, tartrate, and lipoic acid, the functional NPs exhibit effective binding to the positively charged side chains of amino acids in proteins through the negatively charged anionic groups on the NP surface (refer to Fig. 6.1). For instance, gold (Au) NPs synthesized through citrate reduction can be functionalized with carcinoembryonic antibody molecules at pH levels slightly above the isoelectric point of the citrate ligand.

Affinity Interactions: Affinity interaction is highly effective for the bioconjugation of targeting ligands to NPs due to specific and strong complementary recognition interactions, such as antigen–antibody, nucleic acid–DNA, lectin–glycan, streptavidin–biotin, aptamer–protein, aptamer–small biomolecule, and hormone–receptor interactions. Moreover, the presence of multiple binding sites in various biomolecules allows for multidirectional growth of NP structures. For instance, antibodies possess two antigen-binding sites known as Fab (antigen-binding fragment), while streptavidin and concanavalin A each have four binding domains. This characteristic enables the modification of NP surfaces with streptavidin, which specifically binds to biotinylated molecules. The resulting linkage is highly stable and represents the strongest non-covalent interaction. Unlike hydrophobic and electrostatic interactions, affinity binding is not influenced by environmental conditions such as pH changes, salinity variations, or hydrophilicity. Avidin-functionalized Ce/Tb-doped $LaPO_4$ NPs

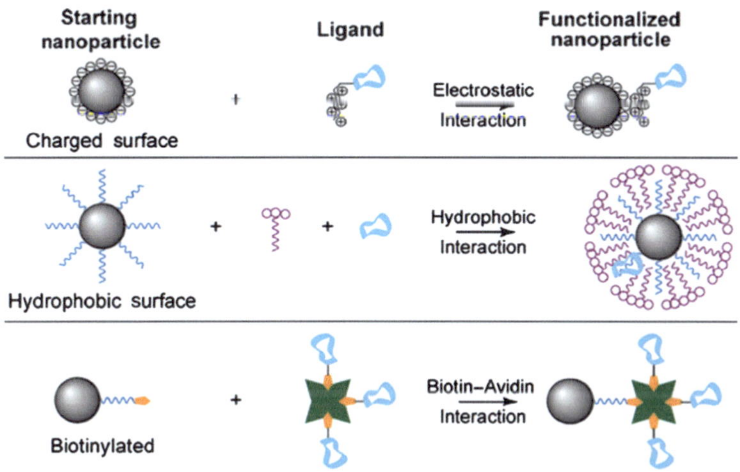

Fig. 6.1 Reaction mechanisms for non-covalent biofunctionalization methods. (Top) Charged nanoparticles interact electrostatically with oppositely charged ligands to form stable conjugates. (Middle) Nanoparticles with hydrophobic surfaces undergo ligand exchange or encapsulation via hydrophobic interactions. (Bottom) Biotinylated nanoparticles bind specifically to avidin- or streptavidin-conjugated ligands through strong biotin–avidin interactions. These approaches enable tailored nanoparticle surface modification for targeted biosensing and biomedical applications [1]

have been utilized for the super assembly of biotinylated proteins and oligonucleotides, demonstrating the practical utility of these bioconjugates in biosensing and biolabeling applications.

Aptamers, short single-stranded nucleic acid sequences obtained through in vitro selection (SELEX), have emerged as promising recognition elements for proteins and other small biomolecules. Aptamers offer several advantages over antibodies, including high specificity and affinity, chemical stability, ready availability, and flexibility. Consequently, aptamers have garnered significant interest as tools for protein detection. Numerous developments have been made in aptamer-based technologies, combining aptamers with techniques such as polymerase chain reaction (PCR), electrochemistry, capillary electrophoresis, mass spectrometry, and quartz crystal microbalance, to achieve highly sensitive protein detection. Aptamer-functionalized NPs present a promising platform for constructing new biosensing and bioanalytical systems by exploiting the specific recognition capability of aptamers toward target biomolecules.

In recent advancements, a nanoprobe consisting of concanavalin A-functionalized gold NPs has been developed based on the specific recognition between lectins and sugar epitopes. This nanoprobe enables highly sensitive and selective in situ evaluation of carbohydrates on living cells through electrochemical tracing of the enzyme co-immobilized on the NPs. Additionally, by coupling efficient cell capture on electrodes modified with arginine-glycine-aspartic acid-serine tetrapeptide-functionalized single-walled carbon nanotubes (SWNTs) through the specific affinity interaction between the tetrapeptides and integrins on the cell surface, dynamic changes in four types of cell surface carbohydrates during drug treatment can be monitored using lectins labeled with horseradish peroxidase (HRP).

Covalent Coupling: In terms of surface functionalization stability and reproducibility, controlled chemisorption through covalent binding is generally preferred over unspecific physisorption. Direct-reaction strategies involve the direct bonding of functional groups on NP surfaces to reactive ligands with the assistance of catalysts. NP surfaces functionalized with sulfhydryl, amine aldehyde, and carboxylic functional groups can be targeted for such reactions.

Firstly, primary binding of thiolated molecules, such as oligopeptides, to gold (Au) NPs allows for the covalent tethering of biomolecules. Similarly, cadmium telluride (CdTe) NPs capped with dimercaptosuccinic acid, which strongly binds to Cd^{2+} ionic sites on the NP surface through its thiol groups, can be functionalized with biomolecules through direct chemical reactions. In some cases, tight chemisorption of proteins on AuNPs can occur when thiol groups from cysteine residues in the proteins bind to the surface of the AuNPs. If native proteins lack thiolated residues, thiol groups can be introduced using 2-iminothiolane through chemical means.

Secondly, NPs decorated with carboxylic acid groups can be covalently bonded to biomolecules containing primary amines using 1-ethyl-3-(3-dimethylaminopropyl) carbodiimide (EDC)/N-hydroxysuccinimide (NHS) linkers. This approach has been used to attach DNA, aptamers, and antibody-antigen complexes to NPs. Multi-walled carbon nanotubes (MWNTs) can be shortened by sonication in a mixture of sulfuric acid (H_2SO_4) and nitric acid (HNO_3) to introduce hydrophilic carboxylate groups for subsequent functionalization (Fig. 6.2). After activation with EDC and NHS, protein molecules like secondary anti-PSA antibodies (Ab_2) can be attached to the surface of MWNTs. Similarly, amino-decorated NPs can conjugate with biomolecules containing carboxylated groups, enabling the attachment of peptides, proteins, antibodies, and enzymes to the NPs.

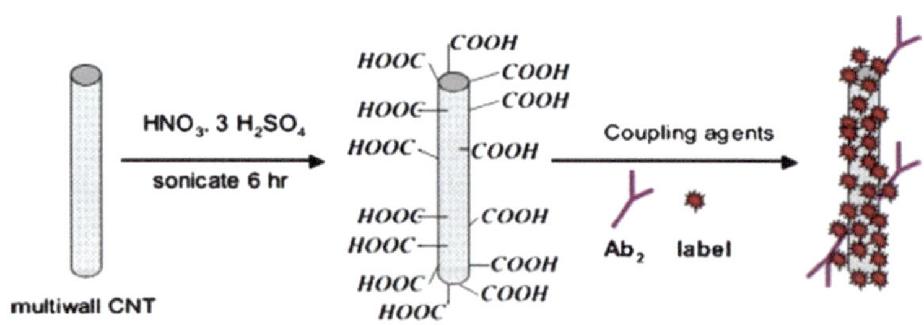

Fig. 6.2 Schematic of covalent coupling and bioconjugation of multiwalled carbon nanotubes (MWCNTs). MWCNTs are first oxidized by treatment with a nitric and sulfuric acid mixture (HNO_3 and H_2SO_4) under sonication for 6 h, introducing carboxylic acid (–COOH) groups on their surface. These functional groups enable covalent attachment of secondary antibodies (Ab_2) and labeling molecules via coupling agents, facilitating their use in biosensing applications through enhanced bioaffinity and signal transduction [2]

Thirdly, the main compounds used for modifying metal oxide NPs include silanes, carboxylates, and phosphonates. Silanes are commonly employed as modifiers for metal oxide surfaces due to their ability to bear various functionalities such as amino, cyano, carboxylic acid, epoxy groups, and more, allowing for subsequent functionalization. Additionally, binding carboxylate ligands and phosphonate groups of biomolecules to metal oxide NPs can be achieved through monodentate coordination, bridging chemisorption, or chelating chemisorption, providing a useful method for biofunctionalizing metal oxide NP surfaces.

When biomolecules, particularly proteins, come into direct contact with an unprotected solid substrate, they can undergo denaturation and lose their specific biochemical functionality. To overcome this issue, low-molecular bifunctional linkers have been widely employed to generate covalently tethered conjugates of biomolecules with various types of nanoparticles (NPs). These linkers possess anchor groups for attachment to the NP surfaces and functional groups for subsequent covalent coupling to the target biomolecules.

Anchor groups like thiols, disulfides, or phosphine ligands are commonly used for binding the bifunctional linkers to NP surfaces such as gold (Au), silver (Ag), cadmium sulfide (CdS), and cadmium selenide (CdSe) NPs. These anchor groups can replace weakly adsorbed molecules, thereby stabilizing the NPs, or they can be incorporated during NP synthesis to create a functionalized NP surface for further reactions.

These reactions enable covalent linkage between functionalized nanocores and biological molecules (proteins, peptides, etc.), facilitating targeted and stable biofunctional assemblies for biosensing and nanomedicine applications [3].

Bifunctional linkers offer a wide range of terminal functional groups. The most frequently used groups include amine, active ester, isothiocyanate, and maleimide. These groups enable covalent coupling of biological compounds through carbodiimide-mediated esterification and amidation reactions or reactions with thiol groups (see Fig. 6.3). The primary role of linker molecules is not only to provide a high density of docking sites for specific biomolecule attachment but also to maintain a sufficiently low density of electronic defects on the NP surface.

A novel method has been developed using phosphorothioate-modified DNA coupled with a short bifunctional fastener (BF) to assemble NPs along DNA strands with precise control over their position and distance. The BF possesses an alkane thiol group at one end for binding to an Au NP and an iodoacetamide group at the other end for binding to a phosphorothioate group on the modified DNA backbone. This approach allows for the precise arrangement of NPs along DNA, offering potential applications in nanoscale assembly and manipulation.

Click reactions are highly advantageous for the conjugation of biomolecules to nanoparticles (NPs) due to their fast and efficient nature, mild reaction conditions, and the formation of water-soluble and biocompatible linkages. Compared to other direct-conjugation strategies, click reactions offer unique features that make them desirable for NP functionalization. Firstly, the azide and alkyne reactive groups used in click reactions exhibit high specificity for one another and do not react with most other functional groups. This ensures precise and specific conjugation at the desired

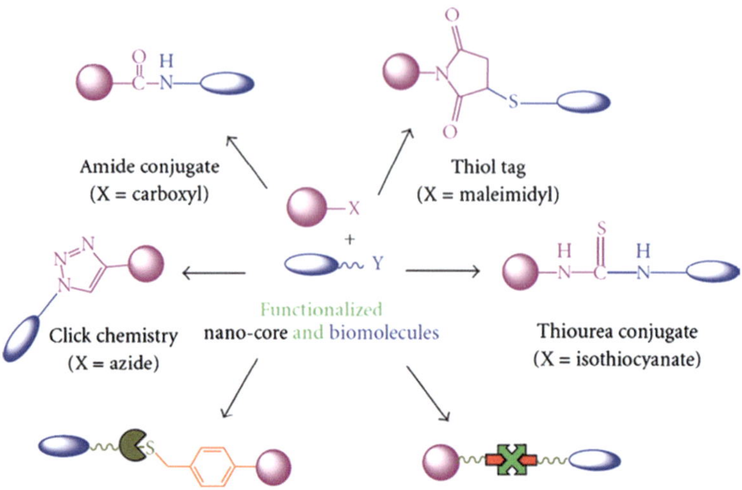

Fig. 6.3 Common chemical strategies for conjugating functionalized nanomaterials with biomolecules. The schematic illustrates key bioconjugation approaches, including amide coupling via carboxyl (X = carboxyl) and amine groups, thiol tagging using maleimide-functionalized groups (X = maleimidyl), thiourea formation with isothiocyanate derivatives (X = isothiocyanate), and click chemistry (X = azide) for efficient and selective azide-alkyne cycloaddition

location on the reactive moiety, avoiding unwanted side reactions. Secondly, click reactions result in highly stable bonds. Unlike amide bonds that can be susceptible to hydrolysis or disulfide linkages that can be cleaved under reducing environments, the bonds formed through click reactions exhibit excellent stability, ensuring the longevity of the conjugated biomolecules on the NP surface. Thirdly, click linkages are characterized by their rigidity, which is beneficial for maintaining the conformation of the reacted moieties at the NP surface. This rigidity prevents undesired cross-interactions and helps preserve the functionality and integrity of the conjugated biomolecules.

Click chemistry methodologies, widely used in polymer and materials science, have also found applications in NP functionalization. NP surfaces can be decorated with either alkyne or azide functionalities, allowing conjugation to complementary functionalized bioactive molecules. Through a simple one-step click reaction, drug-loaded polymer NPs can be functionalized with ligands such as folate, biotin, or even Au NPs, enabling the development of multifunctional NP systems with enhanced targeting or imaging capabilities.

When functionalizing nanoparticle (NP) surfaces with specific groups, such as sulfhydryl, amine, carboxylic, and hydroxyl groups, an activation step is often required to enable these groups to react with the desired ligands. The chosen method for NP surface activation should be compatible with both the ligand and the NPs themselves. Here are some commonly used methods for NP surface activation:

1. Amine Reactive Chemistries: Cyanogen bromide (CNBr) can be used to activate hydroxyl groups by reacting with –OH groups at high pH to produce a highly reactive cyanate ester. This ester then reacts with amine groups. N-hydroxysuccinimide (NHS) esters can also be used to activate –OH and –COOH groups, forming stable amide bonds with primary amines. N,N′-carbonyl diimidazole (CDI) and EDC (1-ethyl-3-(3-dimethylaminopropyl) carbodiimide) are reactive carbonylating reagents that can activate –COOH groups for coupling with primary amine-containing compounds.
2. Aldehyde and Ketone Chemistry: Aldehydes and ketones can react with primary and secondary amino groups to form reversible Schiff bases, which can be reduced and stabilized using reducing agents like sodium borohydride. This method allows the introduction of biomolecules containing primary or secondary amino groups to NPs. Aldehyde groups on the NP surface can be generated by mild oxidation of adjacent diols of agarose with $NaIO_4$.
3. Hydrazide Reactive Chemistry: This method enables the coupling of aldehyde- or ketone-containing ligands through the formation of stable hydrazone linkages. Glycoproteins can be introduced to the NP surface using this method after oxidation with $NaIO_4$ to produce formyl groups on their carbohydrate chains.
4. Sulfhydryl Reactive Chemistry: Iodoacetyl and bromoacetyl activation methods can be used to introduce biofunctional molecules to the NP surface. NPs activated with iodoacetyl or bromoacetyl groups can be coupled to sulfhydryl group-containing ligands through stable thioether bonds. Maleimide is another efficient group for binding sulfhydryl-containing proteins to amine-containing NPs. Cross-linkers like succinimidyl-4-(N-maleimidomethyl)cyclohexane-1-carboxylate (SMCC) have amine-reactive NHS esters on one end and maleimide groups on the other, allowing primary modification of one protein or NP through amine groups and secondary coupling to the sulfhydryl group of another protein.
5. Hydroxyl and Active Hydrogen Reactive Chemistries: Epoxy (bisoxirane) activation, divinylsulfone activation, cyanuric chloride activation, diazonium chemistries, and Mannich condensation are methods used for NP surface activation. Diazonium chemistries involve the use of light to activate aromatic hydrogens or other compounds susceptible to electrophilic attack. Mannich condensation can be used to functionalize NPs with biomolecules containing active hydrogens that can be condensed with formaldehyde and an amine.
6. Photoreactive Cross-Linkers: Photoreactive cross-linkers, such as aryl azide groups, can be used for NP biofunctionalization. These cross-linkers are activated by a light source in a specific range to form reactive nitrenes, which can react with covalent bonds to insert into the structure of a target molecule. Other photoreactive cross-linking reagents include hydrazide derivatives and NHS esters.

6.3 Synthetic Polymeric Materials

The development of supporting films on electrode surfaces for protein or enzyme immobilization has been an active area of research. Among the various polymeric materials used, polystyrene (PS) membrane has gained attention due to its excellent biocompatibility, lack of toxicity, strong adsorption ability, low permeability to small molecules, physical rigidity, and chemical inertness in biological processes. PS membranes are commonly used for enzyme immobilization in enzyme-linked immunosorbent assays (ELISA) through adsorption. However, covalent immobilization of proteins on PS surfaces has been challenging due to the difficulty of introducing functional groups that react with proteins. The multistep methods required for functional group introduction are often tedious and time-consuming.

To address this challenge, polymaleimidostyrene (PMS) has been employed to introduce maleimide groups into the bulk of PS. This approach allows for the coating of enzyme-containing PS membranes onto electrode surfaces under mild conditions, offering new possibilities for biomolecule immobilization.

Interference from electroactive substances and biofouling are two major issues that can affect the performance of biosensors, particularly when electrochemical measurements are conducted in vivo. Biocompatible membranes are desirable as selective barriers and for enhancing biocompatibility in electrochemical biosensors. Cellulose acetate layers have been used as a selective barrier, allowing only small molecules like hydrogen peroxide to reach the electrode while excluding interfering electrochemically active compounds.

Nafion, a perfluorinated sulfonated cation exchanger, has been widely employed as an electrode modifier due to its chemical inertness, thermal stability, mechanical strength, and antifouling properties. Nafion-coated electrodes have been used for the analysis of phenol and the determination of parathion. Nafion is often chosen as a coating material due to its chemical inertness and ease of casting from a solution. The anionic nature of Nafion allows for the formation of hydrophilic channels within a hydrophobic matrix, resulting in films that are robust, exclude anionic interferents, and enhance biocompatibility.

Chitin and chitosan, as functional materials, possess a unique set of characteristics that make them attractive for enzyme immobilization supports. These characteristics include biocompatibility, biodegradability to harmless products, nontoxicity, physiological inertness, antibacterial properties, heavy metal ion chelation, high affinity to proteins, gel-forming properties, hydrophilicity, availability of reactive functional groups for direct enzyme reactions and chemical modifications, mechanical stability and rigidity, and ease of preparation in different geometrical configurations suitable for specific biotransformation processes. Chitin and chitosan are expected to be widely exploited in the future, especially in applications related to enzyme immobilization, due to their remarkable properties and the increasing importance of materials derived from renewable sources. The biodegradability and biocompatibility of chitosan, in particular, make it an attractive green polymer with diverse biological applications.

Conductive polymers, also known as intrinsically conducting polymers (CPs), are organic conjugated polymers with a unique conjugated π-electron backbone system. These polymers have the ability to conduct electricity and have emerged as a promising class of biocompatible materials. The discovery and development of conductive polymers led to Professors Alan Heeger, Alan G. MacDiarmid, and Hideki Shirakawa being awarded the 2000 Nobel Prize in Chemistry. The electronic properties of conductive polymers have been extensively studied, and they have attracted significant scientific and technological interest due to their versatility in synthesis techniques, wide range of properties, and diverse applications in materials science. Previous comprehensive reviews have provided an excellent summary of the earlier work in this field.

Conductive polymers offer high electrical conductivity and optical properties like metals and semiconductors. Additionally, they possess the processability advantages and mechanical properties of polymers, making them suitable for the development of electrochemical biosensors. They can serve as sensitive components, provide a matrix for biomolecule immobilization, signal amplification, and enable rapid electron transfer for the fabrication of efficient biosensor devices. Various monomers can be electropolymerized on an electrode surface to form stable conductive films under appropriate conditions.

Polypyrrole (PPy), polyaniline (PANI), and polythiophene (PT) are examples of conductive polymers that can be obtained by electrochemical polymerization using techniques such as potentiostatic, galvanostatic, or multisweep experiments. The thickness of the polymer films can be controlled by measuring the charge transferred during the electrochemical polymerization process and adjusting parameters like temperature, monomer concentration, polymerization potential or current, as well as the concentration and nature of the supporting electrolyte. Conductive polymer films also exhibit size-exclusion and ion-exchange characteristics, which can reduce the influence of interfering compounds.

The extended conjugated π-electron backbone of conductive polymers, with alternating single and double bonds along the polymer chain, is responsible for their unique electronic properties, including electrical conductivity, low-energy optical transitions, low ionization potential, and high electron affinity. Scientists from various disciplines are now collaborating to study organic solids that exhibit remarkable conductivity. For a polymer to exhibit intrinsic electrical conductivity, there must be an overlap of molecular orbitals to allow the formation of delocalized molecular wave functions. The molecular orbitals should also be partially filled to enable the free movement of electrons throughout the lattice. The electrical conductivity of conductive polymers can vary over several orders of magnitude in response to changes in pH and redox potential in their environment. The electrical properties can be finely tuned through organic synthesis methods and advanced dispersion techniques.

The unique properties of conductive polymers represent a significant advancement in materials research, as they can efficiently transfer electric charge generated by biochemical reactions to electronic circuits. Moreover, conductive polymers can be selectively deposited onto specific areas of electrodes. The study of the distinctive properties of conductive polymers has provided fundamental insights into the chemistry and physics of these novel materials, and they have been successfully applied in the fabrication of amperometric biosensors.

6.4 Inorganic Colloidal Nanoparticles

Colloidal nanoparticle biosensors belong to a class of biological probes that significantly enhance biological sensing capabilities and offer insights into the biomolecular domain. Nanoparticles exhibit unique and valuable characteristics not present in their bulk counterparts, making them exceptional signal transducers for biosensors. For instance, semiconductor nanoparticles like fluorescent quantum dots (QDs) can display bright and tunable fluorescence, while metallic nanoparticles like plasmonic gold nanoparticles (AuNPs) can exhibit localized surface plasmon resonance (LSPR) phenomena. These nanoparticles demonstrate strong responses to stimuli, and their reactions can be controlled by interactions with target analytes, rendering them ideal for biosensing applications.

Nanoparticle biosensors possess the advantage of dynamic interaction with their surroundings, unlike passive labeling techniques employing nanoparticles, dyes, or stains. They can promptly respond to environmental changes in real-time, which is advantageous for both in vitro and in vivo applications. Additionally, nanoparticle biosensors often eliminate the need for washing steps, making them well-suited for swift analytical quantification assays, detecting analytes within cells, and real-time sensing and tracking of analytes in vivo.

Fluorescent quantum dots (QDs) and plasmonic gold nanoparticles (AuNPs) are frequently employed in nanoparticle biosensors. QDs are inorganic semiconductor nanoparticles known for their high fluorescence brightness and photostability. They feature broad absorption and sharp emission spectra that can be finely tuned by particle size, allowing for multiplexing. QDs are efficient energy-transfer agents to nearby donors or acceptors and can be functionalized with various molecules, a crucial aspect of QD biosensors. On the other hand, plasmonic nanoparticles, particularly AuNPs, exhibit unique optical properties due to localized surface plasmon resonance (LSPR). The spatial confinement of surface plasmons in these nanoparticles results in a distinct extinction peak in the visible spectrum. The strong surface fields polarize the local environment around the nanoparticles, enabling interactions with external agents like biomolecules, ions, dyes, or other nanoparticles. This interaction leads to phenomena such as LSPR peak shift, surface-enhanced Raman scattering (SERS), and quenching or enhancement of fluorophores, forming the foundation of biosensors.

While techniques such as enzyme-linked immunosorbent assay (ELISA), polymerase chain reaction (PCR), and fluorescence in situ hybridization (FISH) are widely employed in biomolecular research and diagnostics, they can be labor-intensive and possess limitations. Nanoparticle biosensors offer heightened sensitivity and hold the potential to replace or enhance these existing methods. Overcoming challenges such as reproducibly synthesizing high-quality nanoparticles and optimizing performance in physiological conditions, including unfiltered bodily fluids containing interfering molecules, varying ionic concentrations, and pH fluctuations, is essential for the successful deployment of nanoparticle biosensors. Careful engineering of nanoparticle cores, surfaces, and biosensing mechanisms plays a pivotal role in addressing these challenges and advancing nanoparticle biosensor technology.

For nanoparticle biosensors to be effective, they must exhibit certain characteristics. They need to be chemically robust, capable of withstanding complex conditions, and have minimal impact on the probed system, ideally with low or no toxicity. Furthermore, they should generate strong yet tunable responses to incident light, resulting in a significant signal change upon analyte interaction.

Fluorescent quantum dots (QDs) have been extensively utilized as the foundation for fluorescent biosensors due to their outstanding optical properties. However, concerns about the heavy metal content of typical QDs, often composed of cadmium selenide (CdSe) and zinc sulfide (ZnS), have arisen due to toxicity and disposal challenges. To address these concerns, heavy metal-free alternatives have been developed, incorporating materials like silicon, carbon, ternary I–III–VI alloys (e.g., Cu, Sn, Zn, Ag, In, and S), and conjugated polymers. While biosensing has been demonstrated with these alternatives, none have yet surpassed the optical quality and biofunctionalization capabilities of Cd-based QDs. Nonetheless, ongoing research and development in this area hold the potential to yield heavy metal-free alternatives that outperform Cd-QDs and significantly impact the field of nanoparticle biosensors.

Luminescent upconverting nanoparticles (UCNPs) constitute a recently emerged class of nanoparticles garnering attention for their biosensing applications. Unlike traditional fluorescent nanoparticles that absorb high-energy light and emit at lower energy levels, UCNPs exhibit the reverse behavior. They have the ability to harness low-energy light through multiphoton absorption and convert it into higher-energy emission. This unique property enables UCNPs to operate at red-infrared wavelengths, avoiding issues like autofluorescence and enhancing their penetration capabilities in biological matrices. Ongoing research is focused on enhancing the luminescence brightness and surface chemistry of UCNPs, with a particular emphasis on biosensing applications.

Noble-metal nanoparticles, including gold (Au), silver (Ag), and platinum (Pt), demonstrate distinctive optical properties contingent upon their size and shape. Noble-metal nanoparticles smaller than approximately 2 nm can exhibit tunable fluorescence across ultraviolet, visible, and near-infrared (NIR) wavelengths. They possess biocompatibility, strong fluorescence, prolonged emissive lifetimes, and excellent photostability, rendering them highly appealing for biosensing applications. Their petite size is advantageous, falling below the renal clearance limit of 5.5 nm. Noble-metal nanoparticles exceeding 2 nm in size exhibit localized surface plasmon resonance (LSPR), leading to phenomena like superquenching, fluorescence enhancement, and LSPR field perturbation, all of which can be harnessed for biosensing purposes. The precise engineering of metal alloys or core–shell architectures permits fine-tuning of the plasmonic properties of these nanoparticles. Commonly used alloying elements include copper (Cu), palladium (Pd), and platinum (Pt). Shape manipulation of noble-metal nanoparticles can also shift the LSPR peak position, enabling tailored adjustments for specific applications.

In addition to the core composition of nanoparticles, their synthesis methodologies have a significant impact on their properties. Various nanoparticle characteristics, such as size, shape, crystal structure, defects, dopants, surface morphology, charge, and the density of capping ligands, are highly sensitive to the conditions under which they are synthesized. Solution-phase synthesis, which involves the reaction of precursors in solutions containing coordinating ligands, allows for the production of high-quality nanocrystals with optimal physical and chemical properties. Fine control over nanoparticle structure and morphology can be achieved by adjusting parameters such as stirring rate, vessel morphology, and precursor injection position. However, achieving reproducible synthesis remains a challenge.

To address this challenge, automated robotic systems like the WANDA system at the Lawrence Berkeley National Laboratory, continuous-flow systems, and microfluidic synthesis platforms are being utilized. These approaches offer better control, real-time tuning, and optimization of nanoparticle synthesis, leading to more consistent and reproducible results. Additionally, in situ X-ray irradiation, liquid cell transmission electron microscopy (TEM), and computer simulations are valuable techniques for studying nanoparticle growth and gaining insights into the underlying processes. These methods enable real-time visualization, spectroscopic analysis, and modeling of nanoparticle nucleation, growth, and surface interactions. They contribute to the development of sophisticated synthesis techniques and an improved understanding of nanoparticle characteristics.

6.5 Layer-by-Layer Assembly

The layer-by-layer (LBL) method has attracted significant attention in the field of electrochemistry due to its tremendous potential as a technique for creating nanostructured films with exceptional organization at the nanoscale. These films are fabricated by leveraging specific interactions between organic functional groups, leading to multilayer growth. The resulting films, when deposited onto a solid substrate, exhibit remarkable stability.

An alternative approach involves utilizing the adsorption process of organic molecules, such as polyelectrolytes, through electrostatic attraction, to fabricate films. In most cases, the adsorption processes in the LBL technique rely on the attractive forces between opposite ionic charges found within the molecular structures of polyelectrolytes. To grow multilayer films, the adsorption occurs through an alternating immersion process of a solid substrate (such as quartz, hydrophilic,

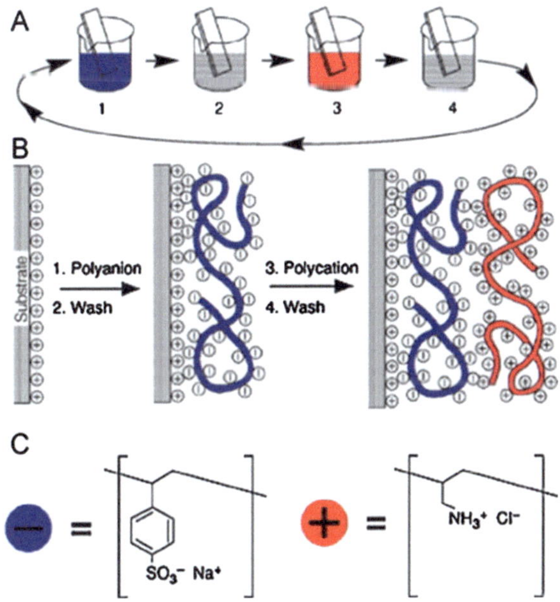

Fig. 6.4 Schematic representation of layer-by-layer (LbL) assembly of polyelectrolyte multilayers. **a** Cyclic deposition process: the substrate is alternately immersed in polyanion (blue) and polycation (red) solutions, with rinsing steps in between to remove unbound material; **b** sequential adsorption of oppositely charged polymers forms alternating layers on the substrate surface via electrostatic interactions; **c** chemical structures of representative polyelectrolytes used: polystyrene sulfonate (PSS^-) as the polyanion and polyallylamine hydrochloride (PAH^+) as the polycation [4]

or hydrophobic glasses) in a solution containing a cationic polyelectrolyte and an anionic polyelectrolyte. Any excess weakly adsorbed polyelectrolyte is subsequently removed by washing with a solvent solution, followed by drying with a flow of nitrogen gas or air (see Fig. 6.4).

Synergistic interactions among the components present on the electrode surface play a crucial role in enhancing the electrochemical properties of biosensors. One of the key advantages lies in the ability to precisely control the characteristics of the film at the molecular level, offering experimental simplicity. A vital aspect of improving biosensor performance is selecting an efficient signal transducer material and establishing effective communication between biomolecules and the electrode surface. This communication is essential for enhancing signal processing on the electrode surface.

While it is widely recognized that nanostructured thin films have the potential to enhance the sensitivity of electrochemical biosensors, significant efforts have been dedicated to creating an environment that mimics biological conditions for biomolecule operation. It is well-established that biomolecules tend to lose their bioactivity when immobilized on solid surfaces due to significant changes in their molecular structure. As mentioned earlier, the self-assembly methodology proves valuable in immobilizing various types of biomolecules and enables effective detection methods.

6.6 Prevention of Non-specific Interactions/Bindings

The primary objective in designing and constructing sensing platforms is to achieve highly precise and accurate measurements of the smallest possible amount of the target analyte in complex real samples. To accomplish this, it is crucial to develop simple and innovative sensor platforms capable of mitigating non-specific binding (NSB) of unwanted proteins or molecules on the electrode surface.

Various strategies have been reported to minimize NSB, and they can be classified into two groups: physical and chemical surface modifications. Physical modification strategies involve attaching molecules directly to the surface, such as blocking buffer solutions, or forming complexes with other particles, for example, surfaces coated with avidin. In physical protein adsorption, forces such as van der Waals forces, hydrophobic interactions, electrostatic interactions, and hydrogen bonding play a role.

On the other hand, chemical modification strategies are more specific compared to physical adsorption. These strategies involve chemical reactions between different residues (such as amine, thiol, and carboxyl) of interacting molecules. Examples of chemical modification strategies include:

(i) Allotropic modification of carbon, such as using carbon nanostructures. (ii) Modification using metal nanoparticles, such as metallic gold/silver nanoparticles and magnetic beads. (iii) Polymerization approaches, including polyethylene glycol (PEG), oligo(ethylene glycol) (oEG), and conducting polymers. (iv) Self-assembled monolayers (SAMs). (v) Diazonium salt surface chemistry. (vi) Sol–gel chemistry modification.

These chemical modification strategies offer diverse ways to tailor the properties of the electrode surface and minimize NSB, ultimately improving the performance of sensing platforms.

One of the extensively studied techniques in biosensor design involves the high-specific binding of oriented antibodies onto electrode surfaces through avidin coating. Avidin is a protein derived from egg white and is composed of four subunits, referred to as tetramers, which can bind one biotin per subunit or four biotin molecules per avidin molecule. The avidin–biotin system exhibits the strongest non-covalent interaction between biomolecules known, with an extraordinary affinity constant of 1015 M−1. Analogues of avidin, such as streptavidin and neutravidin, have been reported to possess similar affinities for biotin molecules and contribute to the improved analytical efficiency of biosensors. Neutravidin, in particular, exhibits the lowest non-specific binding (NSB) among all known binding proteins, including streptavidin.

Another commonly employed strategy to mitigate NSB in biosensors is the use of block buffers. Examples of block buffers include superblock and superblock

T20 in phosphate buffer solution (PBS) and Tris buffer solution (TBS), as well as starting block and starting block T20 in PBS and TBS. Casein and bovine serum albumin (BSA) in PBS and TBS are also frequently utilized as blocking reagents, with BSA being the most commonly used. The ideal blocking reagent should occupy all remaining NSB sites after the adsorption of the coated protein.

A widely adopted technique, known for its simplicity and cost-efficiency in reducing NSB of proteins on bare electrode surfaces, is direct adsorption or passive adsorption of proteins, also referred to as physisorption, forming a biologically active monolayer. Protein physisorption is typically driven by hydrophobic and electrostatic interactions. In general, strong physisorption of proteins has been observed on bare surfaces of carbon paste electrodes (CPE) and screen-printed carbon electrodes (SPCE).

To enhance the biocompatibility of carbon electrode surfaces toward antibodies or proteins and minimize NSB, various functionalization methods are employed. Functionalization is achieved by oxidizing the carbon surface using oxidizing agents such as O_2, N_2O, CO_2, H_2O, and strong chemical solutions like $KMnO_4$, HNO_3, $KClO_3$, H_2SO_4. For example, bare carbon surfaces of SPCE are pretreated by electrochemical methods using a sulfuric acid solution and applying an oxidation potential. Figure 6.5 provides a summary in biosensors' design of the strategies mentioned earlier, including neutravidin-coated SPCE surfaces, blocking reagents, and physisorption for suppressing NSB of proteins.

Prior to protein adsorption, electrode surfaces can be modified to suppress NSB and ensure the proper immobilization of antibodies in an oriented manner, thereby enhancing the sensitivity and selectivity of electrochemical biosensors. Chemical modification strategies include allotropic modification of carbon (e.g., carbon nanostructures), modification using metal nanoparticles (e.g., Au/Ag nanoparticles and magnetic beads), polymerization (e.g., polyethylene glycol (PEG), oligo(ethylene

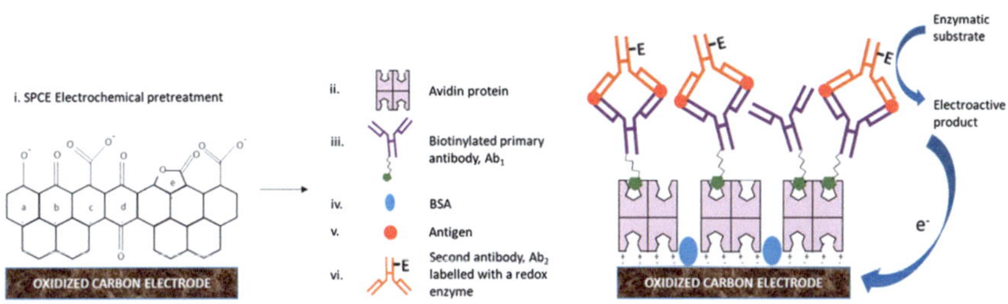

Fig. 6.5 Scheme of biosensor design incorporating strategies to suppress non-specific binding (NSB). (i) Screen-printed carbon electrode (SPCE) is electrochemically pretreated to introduce oxygen-containing functional groups; (ii–iv) avidin is immobilized on the oxidized surface, followed by the attachment of a biotinylated primary antibody (Ab_1) and blocking with bovine serum albumin (BSA) to reduce NSB; (v) the target antigen binds specifically to Ab_1; (vi) a secondary antibody (Ab_2) labeled with a redox enzyme is introduced for signal amplification. Upon addition of an enzymatic substrate, an electroactive product is generated, producing a measurable electrochemical signal [5]

glycol) (oEG), and conducting polymers), self-assembled monolayers (SAMs), diazonium salt surface chemistry, and sol–gel chemistry modification.

References

1. Ju, H., Zhang, X., Wang, J. Biofunctionalization of Nanomaterials. In: NanoBiosensing. Biological and Medical Physics, in *Biomedical Engineering, Springer, New York, NY.* (2011).
2. Yang, H., Xu, W., Liang, X., Yang, Y. & Zhou, Y. Carbon nanotubes in electrochemical, colorimetric, and fluorimetric immunosensors and immunoassays: a review. *Microchim. Acta* **187**, (2020).
3. Kiessling, F., Liu, Z. & Gätjens, J. Advanced nanomaterials in multimodal imaging: Design, functionalization, and biomedical applications. *J. Nanomater.* **2010**, (2010).
4. Richardson JJ, Cui J, Björnmalm M, Braunger JA, Ejima H, C. F. Innovation in Layer-by-Layer Assembly. *Chem. Rev.* **116**, 14828–14867 (2016).
5. Contreras-Naranjo, J. E. & Aguilar, O. Suppressing non-specific binding of proteins onto electrode surfaces in the development of electrochemical immunosensors. *Biosensors* **9**, (2019).

Chapter 7
Transducers: Theory and Applications

Biosensors couple an active bioreceptor with a signal transducer and an electronic amplifier. They utilize biological systems at various levels of integration to specifically recognize analytes. Beyond the biorecognition element, the transducer plays a key role by converting the physicochemical changes induced by the biorecognition reaction into a measurable signal—such as variations in voltage, current, temperature, or optical properties—which is ultimately transformed into a digital output for processing.

- Electrochemical biosensors rely on bioelectrochemical reactions detected using electrodes. Interactions between the sensing surface and analytes—such as enzymatic redox reactions, labeled affinity recognition, or aptamer binding—are measured through amperometric, potentiometric, or conductometric techniques, and converted into qualitative or quantitative electrical signals.
- Optical biosensors represent a broad class of devices based on light–matter interactions involving the biorecognition element. Optical transducers may function through interferometry, fluorescence, luminescence, or colorimetry. This category also includes plasmonic biosensors, which detect changes in plasmonic spectral properties. Such changes occur due to variations in the effective refractive index caused by introducing different materials into the sensor structure.
- Thermal biosensors measure the heat energy released or absorbed in biochemical reactions. In earlier studies on calorimetry, the change in heat was directly monitored to calculate the extent of reaction (for catalysis) or structural dynamics of biomolecules in the dissolved state. However, its exploitation in biosensors led to the development of thermometric and calorimetric devices.
- Mass biosensors encompass devices like quartz crystal microbalance (QCM), containing a microcantilever sensitive to mass variation. In particular, QCM is based on the resonance of surface acoustic waves (SAWs). The principle of SAW technology is based on the piezoelectric properties of the sensors, so as to binding events and their accompanied mass increase at the sensor surface are detectable by a change in the oscillation of the surface acoustic wave.

A. Simonian et al., *Biosensors as Analytical Tools for the 21st Century*,
https://doi.org/10.1007/978-3-032-01726-0_7

7.1 Electrochemical Sensors

Electrochemical transduction methods are the primarily used techniques while developing biosensors mainly due to the easiness of integration and possibility to develop compact biosensors for in situ monitoring. Potentiometric and amperometric enzyme electrodes are well consolidated (after 60 years since Clark's electrode) and widely used as point-of-care (POC) devices. Among all electrochemical transduction methods, it is possible to conceive biosensors based on potentiometry, voltammetry, amperometry, and coulometry.

Potentiometric Sensors

In potentiometric sensors, the potential generated between the two electrodes immersed in the electrolyte solution of interest is measured using a voltmeter (or potentiostat in open circuit voltage mode—about zero current passing through the circuit). It is considered as one of the classical analytical techniques and its applications range include clinical diagnosis, environmental monitoring, industrial processes, and medicine. The device required for the measurement is made without externally polarizing the circuit at about zero currents primarily determined by the potential difference between electrodes. The measurements involve determination of the potential difference between reference electrode and the working electrode. Sometimes two reference electrodes separated by a permselective membrane are used for the measurement of potential. The transducer may be ion-selective electrode if the assembly of the working electrode provides dependence of its potential on the ion sample concentration [1]. The electrodes are electrochemical sensor based on thin films or selective membranes as recognition elements. The technique is based on a simple way to quantify the concentration of ions which is determined by the ion-selective electrodes (ISE) used. The most common ISE electrodes are pH electrodes. Several other ions are (F^-, I^-, CN^-, Na^+, K^+, Ca^{2+}, $NH_4{}^+$).

The formation of the electrode potential is caused by thermodynamically spontaneous reactions directed to leveling the chemical potential of the species on the electrode interface. Thus, for a metal electrode immersed in the solution of a salt, the redox reaction is caused by the difference in the chemical potential of the ion Me^{Z+} in solution and in the points of a metal lattice.

In accordance with ionic equilibrium, some ions can leave the metal lattice and vice versa, some ions can deposit from the solution on the electrode surface. The direction of the reactions depends on the relation between the free energy of the ions in both phases. As a result, the electrode interface undergoes polarization, leading to an electrical potential difference between the electrode and the surrounding medium. This potential can be measured and serves as an indicator of the extent of charge transfer taking place at the interface. The generated potential provides valuable information about the magnitude of charge transfer, allowing us to express it using parameters that describe the nature and quantity of the primary ion involved. This connection is commonly referred to as the Nernst equation.

$$E = E^0 - \frac{2.303RT}{nF} \log\left[\mathrm{Me}^{Z+}\right].$$

Here, E is the electrode potential, E^0 is a standard potential that depends on the nature of the ion and temperature, R is the universal gas constant, T is the Kelvin temperature, Z the charge of the ion, n the number of electrons involved in the redox reaction, and F the Faraday constant. For the thermodynamic behavior of the system, the activity Me of the Me^{Z+} ion corresponds to its concentration. In the opposite case, the activity differs from the concentration by a factor called the activity coefficient.

For a low concentration of the ion, the activity coefficient can be estimated from the ionic strength I, a parameter, calculated from the total ion content in the solution.

In accordance with the Debye–Hückel law,

$$\log \gamma = -0.509\, z^2 \sqrt{I}.$$

The equation can be transformed into a more convenient form for 25 °C.

The value 0.059 V = 59 mV is also called Nernst slope of a calibration curve in the plot of E versus Me.

The reference electrode contains a low soluble salt on the surface of a metal wire. In this case, the potential of the electrode becomes sensitive to the counter ion concentration. If it is high enough, the potential of such an electrode becomes stable and does not depend on the changes in the ionic content of the solution. The reference electrodes are denoted by a metal symbol and an insoluble component on its surface. Two reference electrodes are most frequently used, i.e., the Ag/AgCl electrode and the Hg/Hg_2Cl_2. The internal space of the electrodes can be separated from the external solution by additional filling and porous membranes to avoid internal fouling (double junction ISE). The Nernst equation for an Ag/AgCl electrode is given below:

$$E_e = E^0 - \frac{2.303RT}{F} \log\left[\mathrm{Cl}^-\right] = E^0 - 0.059 \log\left[\mathrm{Cl}^-\right].$$

In electrochemistry, all the potentials are given against a standard reference electrode whose potentials is assumed to be zero. This is a so-called normal hydrogen electrode (NHE). The design of NHE and Ag/AgCl electrodes is presented in Fig. 7.1a, b.

The SHE consists of a Pt plate covered with finely dispersed platinum and immersed in a solution of hydrochloric acid with the activity of hydrogen ions equal to 1. A constant flow of gaseous hydrogen under pressure of 1 atmosphere is bubbled through the solution to saturate the black platinum with hydrogen. Of course, such an assembly is inconvenient for conventional measurements. The potential determined against other reference electrodes can be recalculated to the potential versus SHE in accordance with the simple relations presented in Fig. 7.2.

In aqueous solutions, Ag/AgCl and saturated calomel (SCE) reference electrodes are commercially available and mostly applied in biosensor related investigations. However, both electrodes have limitations. In particular, the Ag/AgCl electrode is

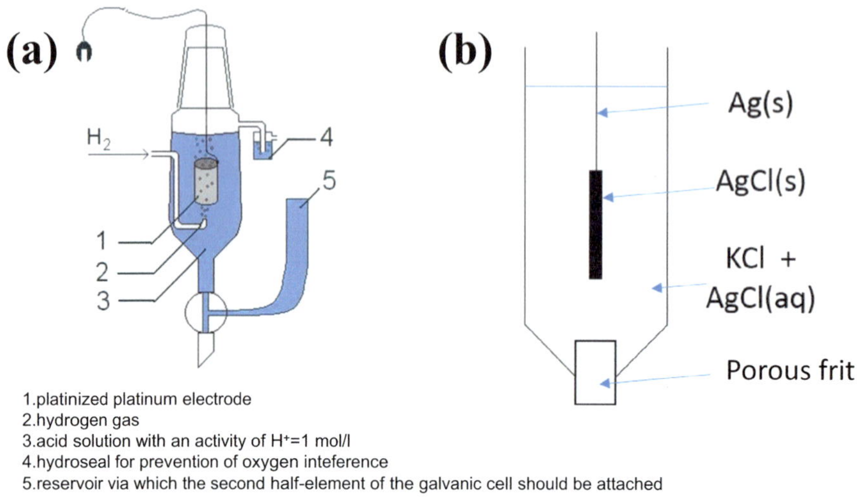

Fig. 7.1 **a** Schematic diagram of a standard hydrogen electrode (SHE). The setup includes: a platinized platinum electrode, bubbling hydrogen gas, an acidic solution with $[H^+] = 1$ mol/L, a hydro seal to prevent oxygen interference, a reservoir for connecting to the second half-cell of a galvanic cell. This configuration is used as a reference electrode with a standard electrode potential of 0 V. **b** Schematic of a silver/silver chloride (Ag/AgCl) reference electrode. It consists of a silver wire (Ag) coated with silver chloride (AgCl), immersed in a solution of KCl saturated with AgCl. The electrode is separated from the test solution by a porous frit, allowing ionic contact while minimizing liquid junction potential

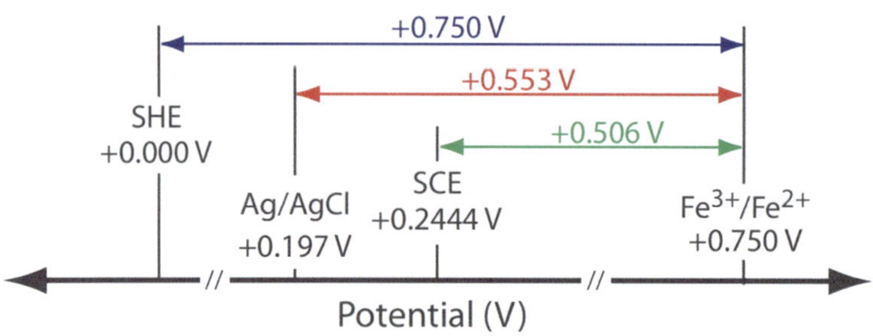

Fig. 7.2 Comparison of electrode potentials for standard reference electrodes and the Fe^{3+}/Fe^{2+} redox couple. The standard hydrogen electrode (SHE) is assigned a potential of 0.000 V. The Ag/AgCl electrode has a potential of + 0.197 V, and the saturated calomel electrode (SCE) has a potential of + 0.2444 V. The Fe^{3+}/Fe^{2+} redox couple lies at + 0.750 V. The arrows indicate the voltage differences between each reference electrode and the Fe^{3+}/Fe^{2+} couple, which are essential for converting measured potentials to the SHE scale

not recommended for use in the media with high amounts of chlorides and other compounds that form insoluble products with Ag^+ ions. The Ag/AgCl electrode is not very stable in water miscible organic solvents. The Hg/HgO electrode is unstable in strong mineral acids. The saturated calomel electrode (SCE) contains toxic mercury and offers some complications in calibration, especially for homemade samples.

Besides reference electrodes, pseudo-reference electrodes have been developed for the application in the assemblies of miniaturized sensors. Thus, the thin film Ag/AgCl electrode can be prepared by lithography and screen-printing techniques by consecutive deposition of the Ag and Ag_2O–AgCl paste on a ceramic or plastic support. Such a design does not imply any internal filling and for this reason does not provide an absolute, but rather a relative stability of the potential in the changing media [2].

Most commercial ion-selective electrodes (ISEs) belong to the family of the electrodes with a plastic membrane and internal filling. The membrane includes an ionophore or an ion exchange. This is a complexing agent responsible for selective binding and the transmembrane transfer of potential determining ion. In addition to the ionophore, the membrane materials contain a plasticizer (e.g., octyl phenyl ether, dioctyl phthalate) and lipophilic salt [3].

The latter is necessary for the acceleration of the transfer of counter ions and reaching the reversible potential. The cation-selective ISEs contain the salt of lyophilic anions and the anion selective electrodes, the salt of lipophilic cation (tetradodecylammonium nitrate). The internal filling contains the salt of the ion to be determined in contact with a metal conductor or an internal reference electrode.

The polymeric membranes of the ISEs described refer to quasi-liquid, or solidified membranes. In miniaturized electrodes made of a glass capillary or a thin plastic cap for pipettes, the membrane can be prepared from a small portion of organic solvent immiscible with water (nitrobenzene) and held in the tip by capillary forces. Such electrodes with very thin tips of several microns in diameter are used as transducers of implantable sensors and biosensors. Their low durability can be compensated for by inert fillers impregnated with membrane components and stuffed in the tip.

Glass ISEs selective for H^+ and Na^+ ions have a similar assembly, but the membrane is made of a special sort of glass that can exchange with the environment by hydrogen and sodium ions. The generation of potential is mainly based on the reversible protonation of the surface silicate groups of glass for pH electrodes and the exchange of Na^+ ions for Na selective ISEs [4].

On the other hand, crystal ion selective membranes are made with poly- or monocrystallines of one substance. For example, a fluoride selective electrode based on $LaBF_3$ crystals exchanges the primary ions between the membrane and the solution. The selectivity of the response is due to the correspondence of the size and charge of the target ion to the vacancies in the crystal structure of the membrane. Besides fluoride sensitive ISEs, crystal membranes are used for the detection of silver cations and halogenide anions (Ag_2O–AgBr or AgI membranes) and sulfides [5].

Solid contact electrodes have been developed to remove the internal filling from the electrode assembly. Instead, the membrane with an ionophore makes contact with special materials that establish electron to ion transduction, i.e., aligning the ion and electron transfer on the interface. Hence, they can convert changes in the ionic content into the electron flow in the metal conductor. Such materials involve various mixtures of the oxides of transient metals that can reversibly oxidize/reduce each other with the simultaneous transfer of an oxygen atom within the layer. The chemical step that appears in an aqueous solution involves the transfer of water molecules or hydrogen ions so that most solid contact electrodes have pH dependent potential irrespective of the ionophore used in the covering membrane.

In the simplest way, the solid contact is obtained by thermal destruction of unstable salts deposited on the electrode surface or by oxidation of the metal electrodes. Some of the electrodes obtained with such a treatment serve for pH measurements. Being less sensitive toward pH than conventional glass electrodes, they can be produced without significant limitations in their sizes and shapes. However, they are used after the deposition of membranes, similarly to traditional ISEs with an internal filling. The membranes can be deposited by drop casting and solvent evaporation, or by immersing the wires into the viscous mixture followed by drying the layer left on the metal surface. In both cases, the ionophore is placed on the material with the ion to electron transduction and specifies the selectivity of the potentiometric response [6–9].

Electroconductive polymers have attracted increasing attention in potentiometric measurements. They are synthesized by chemical or electrochemical oxidation of some aromatic compounds, i.e., aniline, thiophene, and pyrrole and their derivatives. The product of oxidation exerts electron conductivity comparable with that of p-doped semiconductors and even metals. The mechanism of the redox reaction of polyaniline was presented in Chap. 5. In brief, the pH sensitivity of the potential is related to the protonation of amino and imino groups and the involvement of H^+ ions in the reversible redox conversion of imine fragments. As a result, polyaniline has the pH-depending potential in two pH regions. In acidic media, the potential is shifted due to changes in the ratio of reduced and oxidized fragments of the polymer chain as described.

[Ox] and [Red] denote the concentrations of the imino and amino structural units of the polymer, respectively, and n represents the number of H^+ ions and electrons transferred per monomer unit of polyaniline. In basic media, the potential shifts due to changes in the number of charged groups along the polymer chain. Polyaniline, such as in its deprotonated form, is non-conductive in these conditions. Sulfonate made it possible to suppress the pH sensitivity of polyaniline in favor of the potentiometric response toward other ions, in accordance with the ionophore used in the upper polymer layer as shown in Fig. 7.3. On the other hand, the interaction with acidic additives increases the pH sensitivity of polyaniline based sensors [10]. Thus, the addition of the strong water insoluble camphor sulfonic acid to the chemically synthesized polyaniline resulted in an increase of the sensitivity to 89 mV/pH indicating the transfer of one excessive hydrogen ion per monomer unit.

The reaction of protonation and deprotonation of polyaniline is accompanied by the equivalent transfer of counter ions that affect the polymer permeability and the thickness of the surface layer. To some extent, this can be suppressed by the introduction of polyanionic species into the polymer later. Nafion and polystyrol sulfonate

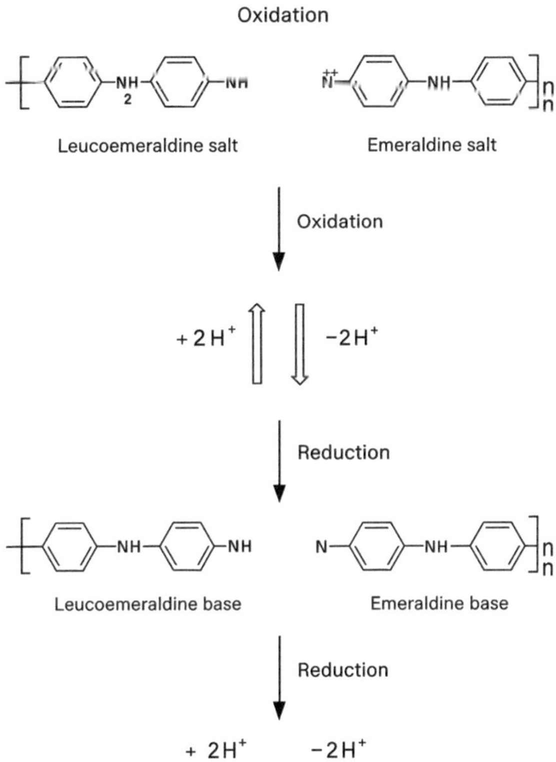

Fig. 7.3 Redox interconversion of polyaniline forms. The diagram illustrates the oxidation and reduction processes between the leucoemeraldine salt, emeraldine salt, leucoemeraldine base, and emeraldine base. Transitions involve proton exchange and changes in oxidation state, highlighting the electrochemical versatility of polyaniline used in sensing, electronic, and electrochromic applications

made it possible to suppress the pH sensitivity of polyaniline resulted in an increase of the sensitivity of 89 mV/pH, indicating the transfer of one excessive hydrogen per monomer unit. The reactions of polypyrrole and polythiophene derivatives do not dramatically differ from those of polyaniline, but the pH dependence of the potential is much wider because the oxidation does not assume the simultaneous transfer of the H^+ ions [11]. As a result, the pG sensors based on polypyrrole and polythiophenes require an additional pH-sensitive ionophore placed on the polymer layer alone or in the PVC matrix. For this purpose, secondary amines with long chain hydrophobic substituents are used.

The solid-state potentiometric sensors based on electroconductive polymers often include mediator systems, e.g., Prussian Blue, hydroquinone derivatives, metal porphyrine complexes, transient metal oxides, and carbon nanotubes that promote the electron exchange within the surface layers and the reversibility of the response. These additives simplify the introduction of other components necessary for the improvement of the sensor performance or biomolecule immobilization as well.

Of special importance is the fact that electropolymerization makes it possible to obtain various nanostructures of the polymers [12]. They can have the shape of tubes, flowers, needles, etc. Their formation increases the internal volume of the surface layer and the specific surface of redox material. The latter does not affect the stationary potential directly but accelerates the achievement of a stationary condition, especially in the microenvironment sharply changed. The nanostructured polymers absorb and confine the biopolymers by multipoint electrostatic interactions in the pores and channels and estimate intimate contact of the biomolecules necessary for a fast and reversible electron exchange.

The formation of the nanostructured polymers can be achieved by the template effect of porous plastic membranes or oxidized aluminum foil fixed on the electrode surface prior to polymerization and removed by appropriate solvents after the polymer deposition. For this purpose, the special regime of electrolysis including several steps that differ in the conditions of the polymer growth has also been developed. Current pulses and the alteration of potentiostatic and galvanostatic electrolysis steps promote formation of the nanostructures.

Besides combinations with conventional plastic membranes bearing ionophores, electropolymerized materials are applied as such for the immobilization of biomolecules and measurement of the changes in the charge distribution and redox potential of the system. Such a non-selective response provides a universal tool for the design of immuno- and DNA-sensors, the reactions of which significantly alter the above parameters on the surface of a transducer. But the signal of such biosensors is preferably recorded in an impedimetric and even amperometric mode than in the potentiometric regime.

The selectivity of the signal of potentiometric sensors is quantified by the potentiometric selectivity coefficients introduced by the Nikolskii–Eisenman equation.

$$E = L_A + \frac{S}{z_A}\log\left(a_A + \sum_i k_{A,I_i} a_{I_i}^{z_A/z_{I_i}}\right),$$

E Measured potential.
L_A Constant for the primary ion (A), without interferences.
S Slope of calibration plot (typically 0.0592 V at 25 °C).
z_A Charge on primary ion A.
a_A Activity of primary ion A.
k_{A,I_i} Selectivity coefficient of analyte A over interfering ion I_i.
a_{I_i} Activity of interfering ion I_i.
z_{I_i} Charge of interfering ion I_i.

The index A belongs to the primary ion and i to the interfering ions. In accordance with, the presence of interfering ions does not affect the slope or the calibration curve but reduces the linear portion, i.e., the concentration range determined by the sensor. There are several methods approved by IUPAC for the estimation of potentiometric selectivity.

The fixed interference method is most frequently used. In them, the e.m.f. is measured for the solution of the constant activity of the interfering ion and the varying activity of the primary ion. The e.m.f. is measured for the solution of the constant activity of the interfering ion and the varying activity of primary ion. The e.m.f. values obtained are plotted against the logarithm of the activity of a primary ion. The intersection of the extrapolated linear portion of the plot yields the A value inserted in the equation.

The matched potential method does not depend on the Nikolskii–Eisenman equation. The potentiometric selectivity coefficient is defined as the ratio of the activities of the primary and interfering ions that give the same potential shift. At first, a known activity of the primary ion i_A added to the reference solution with a fixed activity a_A of the same ion and corresponding shift to the potential is recorded. Then a solution of an interfering ion is added to the reference solution until the same potential shift is reached. The lower the value, the higher the selectivity of the ISE. For a potentiometric biosensor, the calibration graph is linear in semilogarithmic plots with the slope of RT. In the opposite case, the same relationship will be observed for the interfering ion. Although the potentiometric biosensors are mainly based on commercial ISEs with well-known selectivity, in some cases their selectivity has to be taken into account and this is especially true for newly developed potentiometric transducers sensitive to organic species [13, 14]. The application of choline-sensitive electrodes in cholinesterase-based biosensors is limited by a rather high cross sensitivity toward acetylcholine, i.e., a substrate. In the enzyme reaction, the concentration of choline and acetylcholine changes in the opposite direction and the lack of selectivity affects the calibration curve and accuracy of the substrate determination. It is of interest to note that conventional K^+ selective ISEs provide a potentiometric response to choline and can be used in the assembly of choline esterase sensors. The potentiometric selectivity of potentiometric transducers can also impose restrictions on the ionic content of the supporting electrolyte required for potentiometric measurements to maintain the constant ionic strength of the solution.

The whole calibration curve of an ISE in semilogarithmic plots has two extremes and a mid-range linear portion corresponding to the Nernst Equation. The limit of detection is the activity of the primary ion at the point of intersection of the extrapolated linear section and the low concentration level of the signal.

Usually, conventional ISEs make it possible to detect from 10^{-1} to 10^{-6} M of the primary ion. Careful adjustment of the potentiometric membrane content and some modification of the ISE assembly lower the limit of detection (LOD) to the picomolar level. The reproducibility of the e.m.f. measurements is about 1–2%, but semi-logarithmic linearization yields the 7–12% variation of the primary ion concentration. This is significantly higher than the accuracy measurements in voltammetry or spectrophotometric where the signal linearly depends on an analyte concentration. The wide range of linear response and a constant relative deviation of the signal in the whole range of concentrations are the advantages of the potentiometric sensors.

Among the various types of biosensors, potentiometric transducers commonly incorporate pH-sensitive ion-selective electrodes (ISEs) [15]. These biosensors leverage enzymatic reactions such as ester hydrolysis and oxidation of organic matter, which generate organic acids. The quantification of these acids is achieved by detecting the corresponding pH shift in the surface layer. It is important to note that the pH of the bulk solution remains relatively stable due to the minimal conversion of the substrate during a single measurement. Therefore, the pH shift is measured by comparing it against a pH-sensitive transducer without a membrane. This differential signal is independent of chemical reactions that alter the pH or buffer capacity of the solution. This operational mode is particularly suitable for miniature devices like enzyme sensors based on field-effect transistors (FETs) [16]. In this configuration, a pair of FETs is fabricated together on a single support, thereby minimizing complications associated with pseudo-reference electrodes.

It should be mentioned that recording pH changes in an enzymatic reaction is contradictory itself. Indeed, the activity of an enzyme is always pH dependent with the maximum normally within two pH units. For this reason, in a biochemical assay, the enzymatic conversion of a substrate is usually conducted in buffered media to suppress the possible pH shift caused by accumulated products. After immobilization, the support, e.g., gelatin or BSA, exerts its own buffer capacity and is expected to additionally decrease the pH shift. Essentially, the very measures taken to stabilize and replicate the biosensor signal inadvertently diminish the pH shift, which is undesirable for potentiometric transducers.

In the development of biosensors, a compromise is often required between measurement accuracy and signal value. The inclusion of a buffer system is necessary due to the inherent challenges associated with enzyme activity in non-buffered media, which often leads to low reproducibility. Additionally, interferences arising from sample pH fluctuations caused by the accumulation of atmospheric CO_2 can significantly affect the measurements. Furthermore, in acidic solutions, some substrates can undergo hydrolysis. To address these considerations, potentiometric measurements are typically conducted in the presence of millimolar concentrations of a buffer. Moreover, the storage solution used for biosensor maintenance may contain higher quantities of auxiliary reagents aimed at stabilizing the immobilized enzyme between biosensor operations. These measures are implemented to strike a balance between achieving accurate measurements and preserving the integrity of the biosensor system.

The examples of the enzyme sensors based on the ISEs are presented in Table 7.1. Most of the examples refer to pH sensors. The ISEs for ammonia and carbonate ions are also based on pH-sensitive electrodes covered with selective membranes. The determination of choline is the only example of the use of potentiometric sensors specific for organic component [17, 18].

A urea biosensor with enzyme immobilized on to the membrane of the NH_3 selective sensor has been developed for the detection of urea in urine and other biological fluids. The selectivity of the response is achieved by basic media in the internal filling of the electrode. This promotes both the reaction of urease which has the maximum of the pH activity at $pH = 8$, and the transfer of the neutral NH_3 molecules through the membrane. Urease is also compatible enough with polyaniline and polypyrrole and can be immobilized on these polymers without any supports. The hydrolysis of urea results in the formation of two ammonium and one carbonate ion so that the pH sensitivity of the signal is determined by relative acid–base properties of these particles. This reduces the slope of the calibration curve to about 22–30 mV versus theoretical 59 mV/pH for the biosensors based on pH-sensitive transducers [19, 20].

Table 7.1 Reactions involving the release or absorption of ions that may be utilized by potentiometric biosensors

(a) H^+ cation
$D\text{-glucose} + O_2 \xrightarrow{\text{glucose oxidase}} D\text{-glucono-1,5-lactone} + H_2O_2 \rightarrow D\text{-gluconate} + H^+$
$\text{penicillin} \xrightarrow{\text{penicillinase}} \text{penicilloic acid} + H^+$
$H_2NCONH_2 + H_2O + 2H^+ \xrightarrow{\text{urease (pH 6.0)}^a} 2NH_4^+ + CO_2$
$H_2NCONH_2 + 2H_2O \xrightarrow{\text{urease (pH 9.5)}^b} 2NH_3 + HCO_3^- + H^+$
$\text{neutral lipids} + H_2O \xrightarrow{\text{lipase}} \text{glycerol} + \text{fatty acids} + H^+$
(b) NH_4^+ cation
$L\text{-amino acid} + O_2 + H_2O \xrightarrow{L\text{-amino acid oxidase}} \text{keto acid} + NH_4^+ + H_2O_2$
$L\text{-asparagine} + H_2O \xrightarrow{\text{asparaginase}} L\text{-aspartate} + NH_4^+$
$H_2NCONH_2 + 2H_2O + H^+ \xrightarrow{\text{urease (pH 7.5)}} 2NH_4^+ + HCO_3^-$
(c) I^- anion
$H_2O_2 + 2H^+ + 2I^- \xrightarrow{\text{peroxidase}} I_2 + 2H_2O$
(d) CN^- anion
$\text{amygdalin} + 2H_2O \xrightarrow{\beta\text{-glucosidase}} 2\text{glucose} + \text{benzaldehyde} + H^+ + CN^-$

[a] Can also be used in NH_4^+ and CO_2 (gas) potentiometric biosensors
[b] Can also be used in an NH_3 (gas) potentiometric biosensor.es80ll66bp

One potentiometric peroxidase sensor was described. The use of fluoride ISE provides a very selective response toward the primary anion. The target reaction catalyzed by peroxidase includes the oxidation of pentafluorophenol to aquinoid product followed by release of one fluoride ion per substrate molecule. The second approach assumes the detection of the potential related to the reduction of iodine as an artificial electron acceptor involved in the peroxidase reaction instead of molecular oxygen. AH_2 is an organic substrate such as catechol, hydroquinone, ascorbic acid, phenothiazines. Its peroxidase oxidation can be formally ascribed by a transfer of two hydrogen atoms. Appropriate sensors are mainly applied for the detection of hydrogen peroxide but can also be used for detecting peroxidase inhibitors [21, 22].

The first urea biosensor was developed in 1969, seven years after the description of the first amperometric biosensor for glucose determination. For a rather long period of time, the development of potentiometric biosensors matched the amperometric devices stride for stride. Later, however, interest in the potentiometric biosensors lessened due to the drawbacks related to the rather large size of conventional ISEs, the semilogarithmic scale of the calibration plot, and remarkable influence of the matrix component, especially the pH [20, 23].

Voltammetric Sensors: Voltammetry is an electrochemical technique that involves applying a specific voltage profile to a working electrode over time and measuring the resulting current generated by the system. This method allows for the analysis of the electrochemical behavior of chemical species in a solution. By systematically varying the applied potential and monitoring the corresponding current response, valuable information about the redox processes and properties of the studied species can be obtained. Voltammetry is widely used in various scientific disciplines, including analytical chemistry, biochemistry, and materials science, to gain insights into the kinetics and thermodynamics of electron transfer reactions and investigate the behavior of chemical species under different experimental conditions. Voltammetric experiments are carried out in an electrochemical cell consisting of a working electrode, a reference and an auxiliary electrode. As in the case of potentiometric devices, the amperometric biosensor can be a module of all the electrodes assembled in one capsule. However, the biosensor more commonly plays a role of the working electrode. Thus, the measurement is performed in a cell with an analyte solution and two other immersed electrodes. As in potentiometric devices, the working cell contains a constant amount of supporting electrolytes that is necessary for the maintenance of the constant ionic strength and suppression of migration under the electrode polarization. These electrolytes include inorganic salts (KCl, Na_2SO_4) or mineral acids (HCl, H_2SO_4). This technique involves the current at an electrode measured as a function of the potential, applied to the electrode. The working electrode is applied with constant or varied potential and the plot resulting from current versus potential is called voltammogram. The potential applied could be varied in some systematic manner or the scanning of the potential is predetermined by the parameters of the scan rate, initial and final potential.

Voltammetry is applied to analyze any chemical species that is electroactive, which means the species can be made to oxidize or reduce which is determined by the potential of the electrode in a controlled manner. The potential applied causes the species to gain an electron (reduction) or lose an electron (oxidation). Increasing the potential more toward the negative electrode strongly reduces the species and conversely, as the potential becomes more positive it is more strongly oxidizing. The redox reaction occurs on the working electrode and can be controlled by controlling the electrode potential. In modern measurements, the working electrode polarization is established against the auxiliary electrode; however, the potential or voltage is recorded versus reference electrode. This three-electrode setup avoids high currents at reference electrode.

The current that flows is due to the electron transfer which takes place when an oxidation or reduction occurs on the electrode surface. This current, termed "Faradaic," is proportional to the concentration or the current measured is directly related to the number of species oxidized or reduced on the electrode. Here, Ox denotes the oxidized form and Red the reduced form of the electroactive compound.

$$\mathrm{Ox} + ne^- \rightarrow \mathrm{Red}$$

$$\text{Red}-ne^- \rightarrow \text{Ox}.$$

The reaction shown above occurs at different electrodes one on the working electrode and the other on the counter electrode. The reduction current is called cathodic current and is given positive sign, and the current due to oxidation is anodic current and is given a negative sign.

In addition to electron transfer, other electrode reactions of organic species can include chemical steps, e.g., hydrogen/oxygen transfer, dimerization, cyclization, etc.

The general rules of electrochemistry.

Rule 1. The first rule is the requirement of the closed electrical circuit. This means that at least two electrodes must be present in the electrochemical cell. From a purely electrical point of view, it means that we have a sensor electrode (the working electrode) and a signal return electrode (often called auxiliary electrode).

Rule 2. The second rule is the condition of electroneutrality. It means that in an electrochemical cell, the sum of positive charges must equal the sum of negative charges. Thus, separation of positive and negative charges occurs at every interface, but their sum is always zero.

Rule 3. It is not so much a rule as it is an important general point regarding the nature of the interfacial reaction in all electrochemical sensors. Charge transport within the transducer part of the sensor, and/or inside the supporting instrumentation, is electronic. On the other hand, the charge transport in the sample can be electronic, ionic, or mixed (electronic/ionic). An electron transfer at the electrode sample interface and its mechanism becomes one of the most critical aspects of sensor performance.

Faraday's law; the chemical change is caused by the passage of charge Q through the electrochemical interface, at which the electronic conduction in the metal changes into ionic conduction in the sample.

$$m = \frac{Q}{nF} \text{ and } F = e \times \text{NA},$$

where the symbol F is the Faraday constant (96,493 C), n is the number of electrons, NA is the Avogadro constant, and m is the number of moles converted in a reaction.

$$O + ne \leftrightarrow R,$$

where O represents the oxidized species, R the reduced species, and O and R are a redox couple. Species O and R differ only by n electrons. The fundamental relationship that governs the types of behaviors in oxidation and reduction is the Nernst equation, which, for the reaction:

$$O + ne \leftrightarrow R$$

$$E = E^0 + \frac{RT}{nF}\ln\left(\frac{[\text{Ox}]}{[\text{Red}]}\right),$$

where E is the potential of an electrode versus a reference electrode,

E^0 is the standard potential,
R is the gas law constant, 8.31441 J mol^{-1} K^{-1},
T is the absolute temperature, and
F is Faraday constant, 9.648×10^4 C mol^{-1}.

Measurement of Current: Controlled-potential techniques: Study of electron transfer processes at the electrode-solution interface. Based on dynamic (no zero current) situations. The controllable parameter forces the chemical species to gain or lose electrons, resulting in generation of current (units-Ampere). Measure any chemical species that are electroactive (Redox reactions).

Analytical techniques of electrolytic cells: **Amperometry**: Set E_{applied} so that desired reaction occurs, Stir solution, **Measure Current**.

Voltammetry: Quiet or stirred solution, Vary ("scan") E_{applied}, **Measure Current**. Indicates reaction rate. Reaction at electrode surface produces concentration gradient with bulk solution. Mass transport brings unreacted species to electrode surface.

The reference electrode R is placed as close as possible to the working electrode W and is connected to the instrument through a high-resistance circuit that draws no current from it. Because the flow cannot occur through the reference electrode, a current-carrying auxiliary (counter) electrode C is placed in the solution to complete the current path. The current flows through the solution between the working and the auxiliary electrodes. Symmetry in the placement of electrodes is important for the assumption that the current paths from all points on the working electrode are equivalent (Fig. 7.4).

Basic Principle: When species are oxidized or reduced at an electrode, the current produced is directly related to the concentration of the species. The potential of a solution containing the reduced form of a couple **Red** and the oxidized form **Ox** is easiest understood in terms of the Nernst equation as follows:

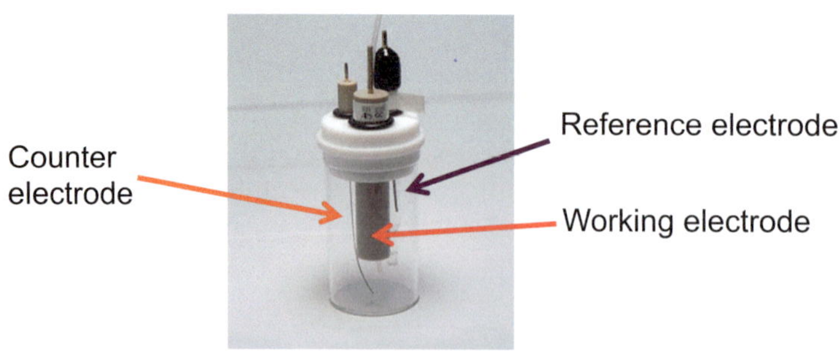

Fig. 7.4 Three electrodes setup

$$E = E^0 + \frac{RT}{nF} \ln\left(\frac{[\mathrm{Ox}]}{[\mathrm{Red}]}\right).$$

The formal potential (E^0) is a parameter characteristic of a couple. Let's look at the example of ferrocene/ferricinium couple.

$$\mathrm{fc}^+ + \mathrm{e}^- \rightleftharpoons \mathrm{fc} \quad \text{at } E_1^0.$$

A widely studied couple is the ferrocene (fc)/ferricinium (fc^+) couple, where there is a single electron transfer involved: This system, along with a series of derivatives, is popular because both the reduced form (ferrocene) and the oxidized form (ferricinium) are both reasonably stable in solution, which means that they can be readily inter-converted. If a solution is prepared consisting of 0.01 mol/dm^3 ferrocene (fc) and 0.01 mol/dm^3 ferricinium ion (fc^+) in acetonitrile, the potential of this solution according to the Nernst equation would be:

$$E = E^0 + \frac{RT}{F} \ln\left(\frac{[\mathrm{Fc}^+]}{[\mathrm{Fc}]}\right) = E_1^0.$$

In order to understand the applications of amperometry, it is necessary to briefly discuss what happens when an electrode is placed in an electrolyte solution. Electric double layer, mass transport, diffusion, Faradaic processes, non-faradaic processes.

Electric Double Layer

Heterogeneous System: Electrode/solution interface. The electrical double layer, *e*'s in electrode; ions in solution—important for voltammetry: Compact inner layer: d_0 to d_1, E decreases linearly. Diffuse layer: d_1 to d_2, E decreases exponentially, as shown in Fig. 7.15.

Electrolysis: Faradaic and Non-Faradaic Currents: Two types of processes at electrode/solution interface that produce current. Direct transfer of electrons, oxidation or reduction. Faradaic Processes. Chemical reaction rate at electrode proportional to the Faradaic current.

Non-Faradaic Current: Due to change in double layer when E is changed; not useful for analysis. Mass Transport: continuously brings reactant from the bulk of solution to electrode surface to be oxidized or reduced (Faradaic). Convection: stirring or flowing solution. Migration: electrostatic attraction of ion to electrode. Diffusion: due to concentration gradient.

Faradaic Current: The current resulting from a change in oxidation state of the electroactive species is termed the ***faradic current*** because it obeys Faraday's law (i.e., the reaction of 1 mol of substance involves a change of $n \times 96{,}487$ C). The faradic current is a direct measure of the rate of the redox reaction. The resulting current–potential plot, known as the ***voltammogram***, is a display of current signal (vertical

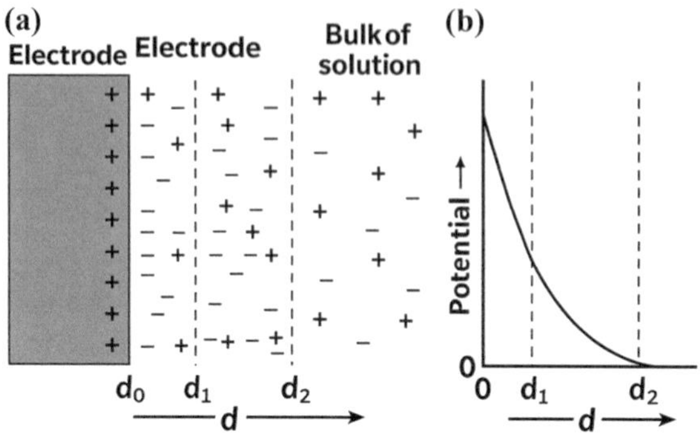

Fig. 7.5 Illustration of the electrical double-layer structure at the electrode–electrolyte interface. **a** Distribution of ions near the charged electrode surface, showing the compact layer (d_0–d_1) and the diffuse layer (d_1–d_2) extending into the bulk solution. **b** Corresponding potential profile, indicating a sharp potential drop near the electrode followed by a gradual decay into the bulk solution [24]

axis) versus the excitation potential (horizontal axis). The exact shape and magnitude of the voltammetric response are governed by the processes involved in the electrode reaction. The total current is the summation of the faradic currents for the sample and blank solutions, as well as the non-faradic charging background current. The pathway of the electrode reaction can be quite complicated and takes place in a sequence that involves several steps. ***The rate of such reactions is determined by the slowest step in the sequence***.

Simple reactions involve only mass transport of the electroactive species to the electrode surface, the electron transfer across the interface, and the transport of the product back to the bulk solution.

More complex reactions include additional chemical and surface processes that precede or follow the actual electron transfer.

The net rate of the reaction, and the measured current, ***may be limited*** either *by mass transport* of the reactant or by *rate of electron transfer*.

The more sluggish process will be a rate-determining step.

Mass Transport-Controlled Reactions

Diffusion—*The spontaneous movement* under the influence of *concentration gradient* (i.e., from regions of high concentration to regions of lower concentration), aimed at minimizing concentration differences.

Migration—*Movement* of charged particles along an electrical field (i.e., the charge is carried through the solution by ions according to their transference number).

Convection—*Transport* to the electrode by a gross physical movement; such fluid flow *occurs with stirring or flow of the solution* and with rotation or vibration of the electrode (i.e., forced convection) or due to density gradients (i.e., natural convection) as shown in Fig. 7.6.

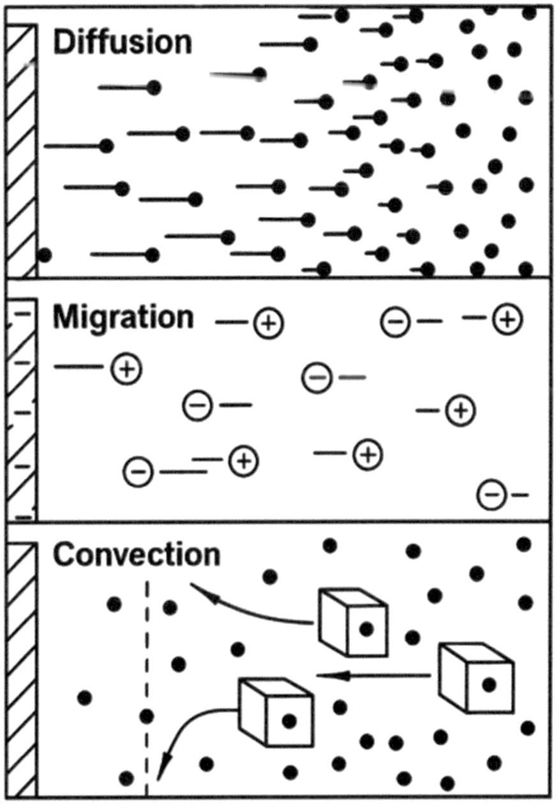

Fig. 7.6 Mass transport controlled reactions: diffusion, migration, and convection

Diffusion Equations for Electroanalytical Processes: The reaction occurring at the surface of the electrode generates a concentration gradient adjacent to the surface, which in turn gives rise to a diffusional flux. According to Fick's first law, the rate of diffusion (i.e., the flux) is directly proportional to the slope of the concentration gradient:

$$J(x,t) = -D\frac{\mathrm{d}C(x,t)}{\mathrm{d}x}.$$

The current (at any time) is proportional to the concentration gradient of the electroactive species:

$$i(t) = nFAD\frac{\mathrm{d}C(x,t)}{\mathrm{d}x}.$$

Nernst–Planck Equation: The flux (J) is a common measure of the rate of mass transport at a fixed point. It is defined as the number of molecules penetrating a unit area of an imaginary plane in a unit of time and has the units of mol cm^{-2} s^{-1}. The flux to the electrode is described mathematically by a differential equation, known as the Nernst–Planck equation, given here for one dimension:

$$J(x,t) = -D\frac{\mathrm{d}C(x,t)}{\mathrm{d}x} - \frac{zFD}{RT}C\frac{\mathrm{d}\phi(x,t)}{\mathrm{d}x} + C(x,t)v(x,t),$$

where D is the diffusion coefficient, $\frac{\mathrm{d}C(x,t)}{\mathrm{d}x}$ is the concentration gradient, $\frac{\mathrm{d}\Phi(x,t)}{\mathrm{d}x}$ is the potential gradient, z and C are the charge and concentration, respectively, of the electroactive species, and $V(x,t)$ is the hydrodynamic velocity. In aqueous media, D usually ranges between 10^{-5} and 10^{-6} cm^{-2} s^{-1}. The current i is directly proportional to the flux:

$$i(t) = -nFAJ.$$

The situation is quite complex when the three modes of mass transport occur simultaneously. This complication makes it difficult to *relate the current to the analyte concentration.*

The situation can be greatly simplified by suppressing the electro-migration or convection, through the **addition of excess inert salt** or use of an inactive solution, respectively. Under these conditions, the movement of the electroactive species is limited by diffusion.

Potential-Step Experiment: A potential-step experiment involving *the reduction of O to R.*

A potential value corresponding to complete reduction of O in a quiescent solution is applied at a planar electrode imbedded in a planar insulator. (Only O is initially present in solution.) Since the surface concentration of O is zero at the new potential, a concentration gradient is established near the surface. The region within which the solution is depleted of O is known as the ***diffusion layer***, and its thickness is given by δ.

Cottrell Equation: Initial and boundary conditions in such experiment include: (1) $C_0(x, 0) = C_0(b)$ (i.e., at $t = 0$, the concentration is uniform throughout the system and equal to the bulk concentration); (2) $C_0(b)$, $C(0, t) = 0$ for $t > 0$ (i.e., at later times the surface concentration is zero); and $C_0(x, 0)$ $C_0(b)$ as x ∞ (i.e., the concentration increases as the distance from the electrode increases). Solution of Fick's laws (for linear diffusion, i.e., a planar electrode) leads to the well-known ***Cottrell Equation*** (Fig. 7.7). Therefore, in quiescent (unstirred) solution and under the condition of linear diffusion, the *mass transport coefficient* decreases with the square root of time when the potential is applied as a single step to the electrode. The diffusion limiting current then also decreases with $t^{1/2}$.

$$i(t) = \frac{nFAD_O C_O(b)}{\sqrt{\pi D_O t}}.$$

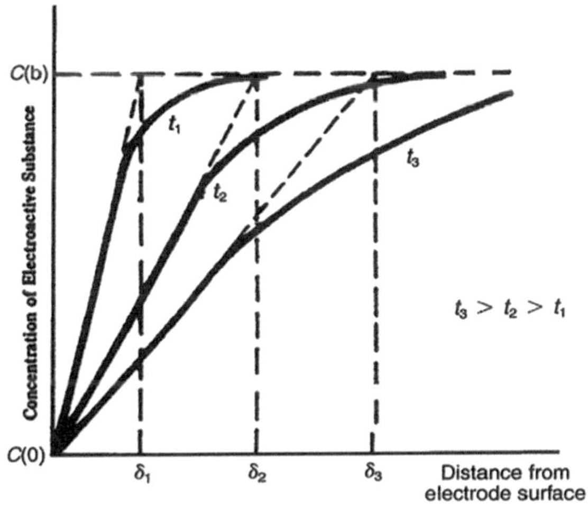

Fig. 7.7 Principle of Cottrell equation: time-dependent concentration profiles of an electroactive species near the electrode surface. Curves labeled t_1, t_2, and t_3 correspond to increasing times ($t_3 > t_2 > t_1$). At each time point, the diffusion layer thickness (δ) increases, indicating the progressive spread of the concentration gradient from the electrode surface $C(0)$ toward the bulk solution concentration $C(b)$. This illustrates how mass transport by diffusion evolves with time during an electrochemical process

Potential-Sweep Experiments: A voltammetric experiment involving: a linear potential scan, the reduction of O to R in a quiescent solution. Linear sweep voltammogram and typical diffusional voltammogram in stirred solutions are shown in Figs. 7.8 and 7.9, respectively.

The thickness of the stagnant (diffusion) layer δ is inversely proportional to the stirring rate U of the solution near the electrode. So, as it becomes thinner, the limiting current i_L increases with the stirring rate, where U is stirring rate, B and α are constants (Figs. 7.10 and 7.11).

Linear Sweep and Cyclic Voltammetry: Cyclic voltammetry consists of scanning linearly the potential of a stationary working electrode (in an unstirred solution), using a triangular potential waveform. Single or multiple cycles can be used. During the potential sweep, the potentiostat measures the current resulting from the applied potential. The resulting plot of current versus potential is termed a cyclic voltammogram as shown in Fig. 7.12.

In cyclic voltammetry, only the oxidized form O is present initially. A negative-going potential scan is chosen for the first half cycle, starting from a value where no reduction occurs. As the applied potential approaches the characteristic E^0 for the redox process, a cathodic current begins to increase, until a peak is reached. After traversing the potential region in which the reduction process takes place (at least $90/n$ mV beyond the peak), the direction of the potential sweep is reversed. During

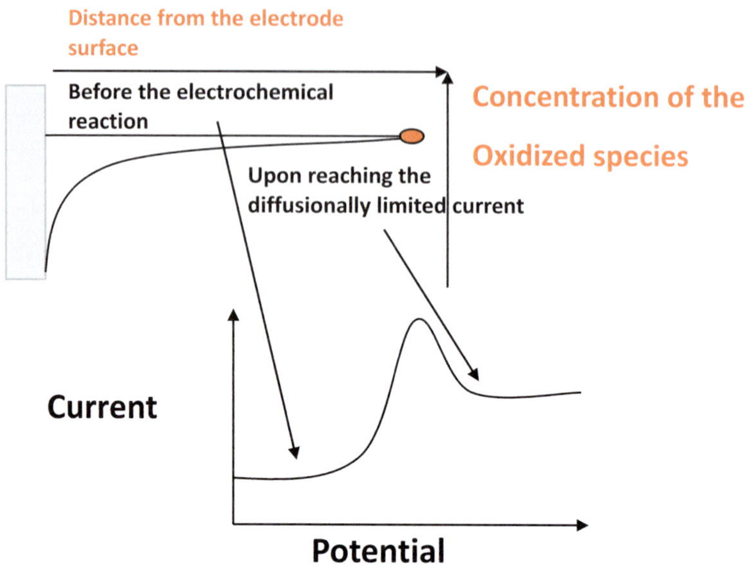

Fig. 7.8 Linear sweep voltammogram and the concentration profile

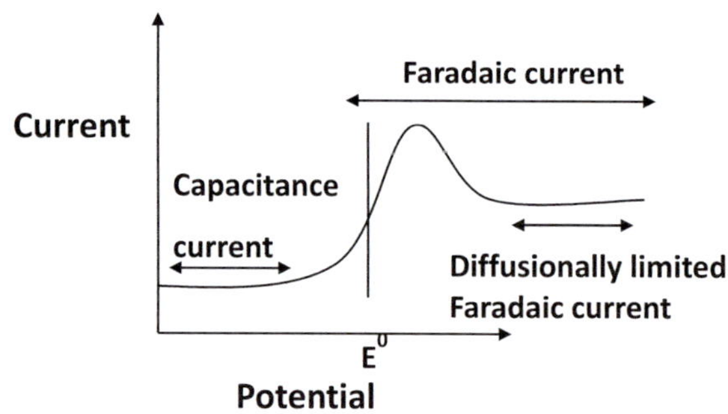

Fig. 7.9 Typical diffusional voltammogram in stirred solutions

the reverse scan, *R* molecules (generated in the forward half cycle and accumulated near the surface) are re-oxidized back to *O* and an anodic peak results (Fig. 7.13).

What do the potentials of the CV peaks tell us? For the reversible electron transfer situation, the peak current (i_p) is given by the Randles–Ševčík equation:

$$i_p = 2.69 \times 10^5 \cdot n^{3/2} A \sqrt{D}\, C_{Ox} \sqrt{v}.$$

The position of both peaks (the anodic peak, E_{pa} and the cathodic peak, E_{pc}) is related to the formal potential of the redox process.

The formal potential for a reversible couple is centered between E_{pa} and E_{pc}:

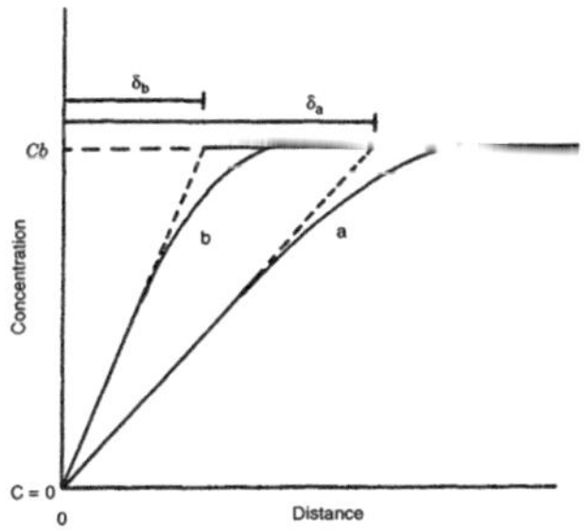

$$\delta = \frac{B}{U^{\alpha}}$$

$$i_l = \frac{nFAD_O C_O(b,t)}{\delta}$$

Fig. 7.10 Comparison of concentration profiles and diffusion layer thicknesses under different mass transport conditions. Curve a represents slower transport with a thicker diffusion layer (δ_a), while curve b corresponds to enhanced transport (e.g., by stirring or convection), resulting in a thinner diffusion layer (δ_b). In both cases, the concentration increases from zero at the electrode surface to the bulk concentration (C_b) at a certain distance. This illustrates how hydrodynamic conditions affect the rate of mass transport to the electrode

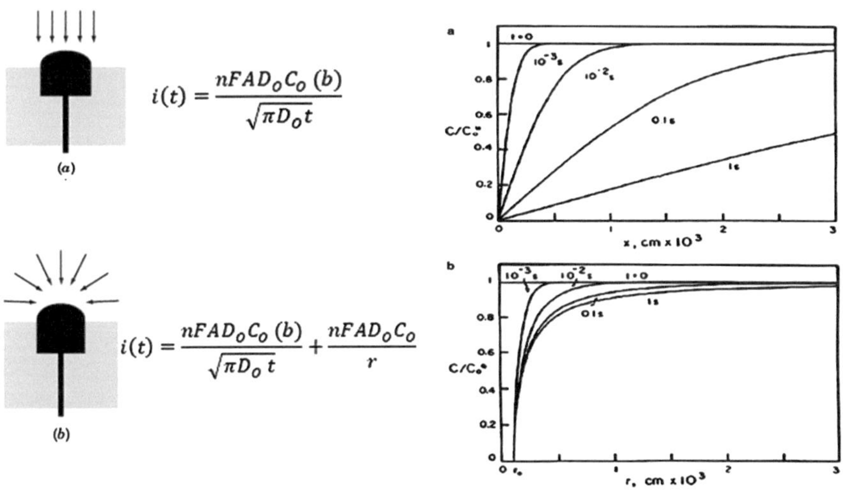

$$i(t) = \frac{nFAD_O C_O(b)}{\sqrt{\pi D_O t}}$$

$$i(t) = \frac{nFAD_O C_O(b)}{\sqrt{\pi D_O t}} + \frac{nFAD_O C_O}{r}$$

Fig. 7.11 Voltammograms: S-shape or peak shape, planar and hemispherical electrodes. Normalized concentration profiles (C/C_0) versus distance* for various diffusion times (e.g., 10^{-3}, 10^{-2}, 0.1, 1 s). It is labeled as subfigures **a** and **b**

$$E^{0\prime} = \frac{E_{pa} + E_{pc}}{2}.$$

The separation between the peak potentials (for a reversible couple) is given by:

$$\Delta E_p = E_{pa} - E_{pc} = \frac{59}{n}\,\text{mV},$$

n is number of electrons.

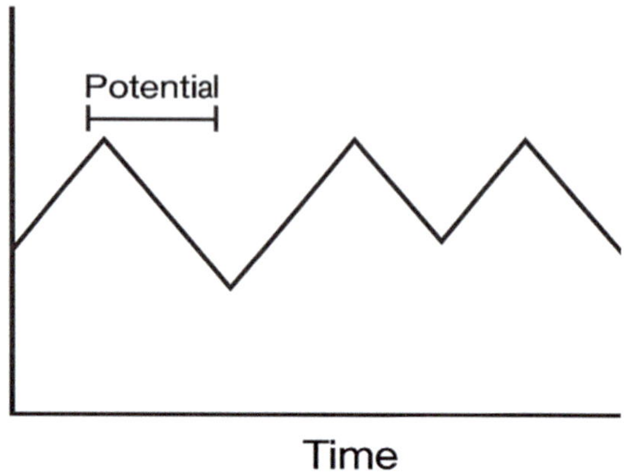

Fig. 7.12 Schematic representation of a triangular waveform commonly used in cyclic voltammetry. The potential is linearly swept back and forth over time, allowing the study of redox processes at the electrode surface by repeatedly cycling the potential through oxidation and reduction regions

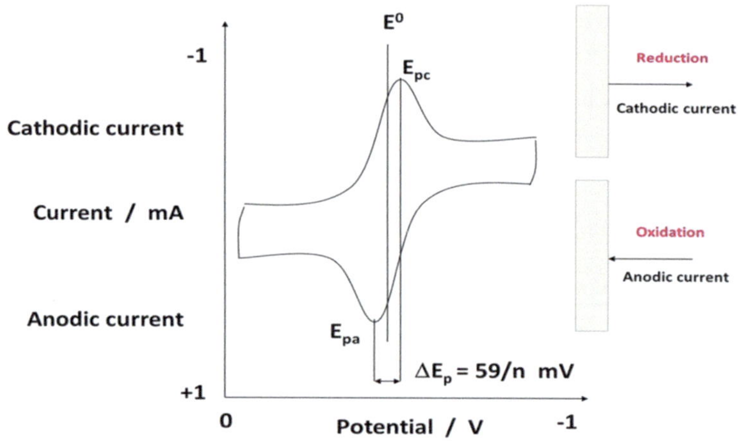

Fig. 7.13 Cyclic voltammogram illustrating a reversible redox couple. The plot shows current (mA) versus applied potential (V), with cathodic and anodic peaks corresponding to reduction and oxidation processes, respectively. E_{pc} and E_{pa} represent the cathodic and anodic peak potentials, while E^0 is the formal potential. The peak separation $\Delta E_p = E_{pa} - E_{pc}$ is ideally 59/*n* mV for a reversible one-electron transfer. The direction of electron flow during oxidation and reduction is indicated alongside

How to recognize diffusional electrochemical processes: To prove the diffusional character of liner sweep or cyclic voltammogram, we need to perform the measurements with different potential scan rates. The linear relationship between current and square root of scan rate potential fulfills the diffusional electrochemical process (Fig. 7.14).

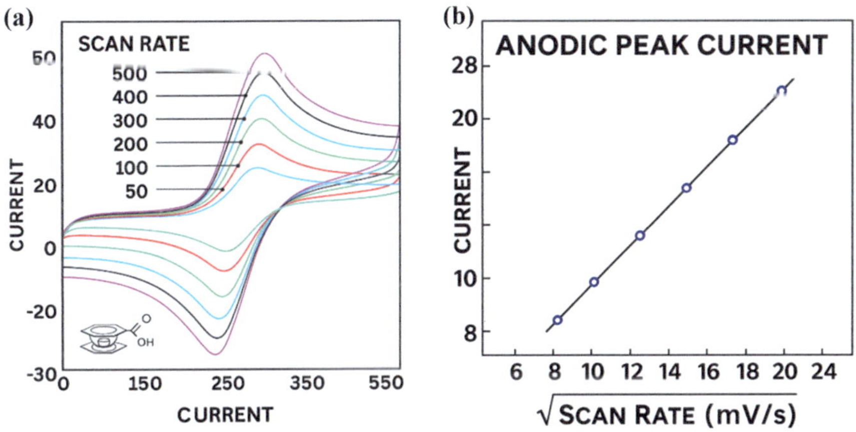

Fig. 7.14 **a** Cyclic voltammograms recorded at various scan rates (50–500 mV/s) showing the redox behavior of the analyte (likely ferrocene derivative as depicted). The increasing peak currents with scan rate indicate diffusion-controlled electrochemical behavior. **b** Plot of anodic peak current versus the square root of scan rate. The linear relationship confirms that the redox process is diffusion-controlled, consistent with the Randles–Ševčík equation for reversible systems

A redox couple fulfilling all the given criteria below will probably be reversible. The redox couple is almost certainly not reversible if one or more of these criteria are not fulfilled

- $I_{pc} = I_{pa}$ (p_c, peak cathodic; p_a, peak anodic).
- The peak potentials, E_{pc} and E_{pa}, are independent of the scan rate v.
- $E^{0'}$ is positioned midway between E_{pc} and E_{pa}, so $E^{0'} = (E_{pa} + E_{pc})/2$.
- I_p is proportional to $v^{1/2}$.
- The separation between E_{pc} and E_{pa} is 59 mV/n for an n-electron couple.

Effect of the electron transfer rate constant on the shape of linear sweep or cyclic voltammograms. To calculate the rate constant we need to know: (1) ΔE as a function of the potential scan rate. (2) Number of electrons per molecule. (3) Diffusion coefficient and concentration of the redox species.

Quasi-Reversible Diffusional CV

In the study of reaction mechanisms, a chemical reaction precede the redox process denoted by letter E and a chemical reaction succeed the redox process by letter C. For the redox and chemical steps, respectively, in the order of the steps in the reaction scheme (Fig. 7.15).

An example of EC mechanism:

$$\text{Ox} + ne^- \rightleftarrows R.$$

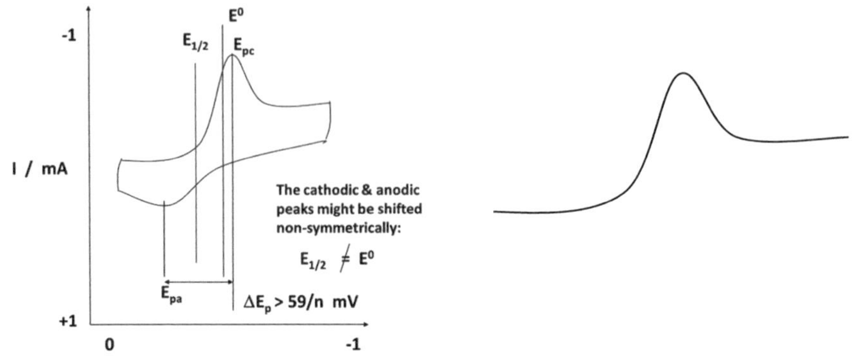

$$Q + e^- \rightleftharpoons Q^- \text{ reversible electron transfer step}$$

$$Q^- + H^+ \rightleftharpoons QH \text{ irreversible chamical step (protonation)}$$

Fig. 7.15 Cyclic voltammogram depicting an irreversible or quasi-reversible redox process. The anodic (E_{pa}) and cathodic (E_{pc}) peak potentials are asymmetrically shifted, resulting in a peak separation (ΔE_p) greater than the ideal 59/n mV. The half-wave potential $E_{1/2}E_{1/2}$, calculated as the average of E_{pa} and E_{pc}, no longer corresponds to the formal potential E^0E^0, indicating deviation from reversibility

For EC mechanism:

$$\text{Ox} + ne^- \rightleftarrows R \rightarrow Z.$$

The cyclic voltammogram will exhibit a smaller reverse peak, because the product R is chemically removed from the surface. The peak ratio i_{pr}/i_{pf} will be smaller than unity; the exact value of the peak ratio can be used to estimate the rate constant of the chemical step. In the extreme case, the chemical reaction may be so fast that all of R will be converted to Z, and no reverse peak will be observed. Additional information on the rates of coupled chemical reactions can be achieved by changing the scan rate.

An irreversible (quasi-reversible) process can be reversible: (1) If the process is limited by the slow electron transfer, make the scan rate slower. (2) If there is an irreversible chemical step after the fast electron transfer process: make the scan rate faster.

Study of Adsorption Processes—Surface Confined.

CV can also be used for evaluating the interfacial behavior of electroactive compounds. Both the reactant and the product can be involved in an adsorption–desorption process. A gradual increase of the cathodic and anodic peak currents is observed, indicating progressive adsorptive accumulation at the surface. The separation between the peak potentials is smaller than expected for solution-phase processes. Ideal Nernstian behavior of surface-confined non-reacting species is manifested by symmetrical cyclic voltammetric peaks. Figure 7.16a shows a cyclic voltammograms for 1×10^{-6} M riboflavin in a 1 mM sodium hydroxide solution. Ideal cyclic voltammetric behavior for a surface layer on an electrode is shown in Fig. 7.16b. The surface coverage, *T*, can be obtained from the area under the peak. The **peak current** is directly proportional to the surface coverage *T* and potential ***scan rate***: In practice, the ideal behavior is approached for relatively slow scan rates and for an adsorbed layer that shows no intermolecular interactions and fast electron transfers [25–27].

Oxygen Electrodes

General aspects of amperometric sensing are illustrated on one of the most ubiquitous and important species, **oxygen**. Amperometric sensors for oxygen have applications extend over many fields: from biology and medicine to industrial applications, energy production, and safety.

Amperometric Sensors

A more fruitful approach to higher selectivity is to increase the difference between the total electrode resistance to the interferants, relative to the resistance belonging to the analyte. The rate constant of the desired charge-transfer reaction can be increased by electrocatalytic lowering of charge-transfer resistance. For example, incorporation of a highly selective oxidation catalyst such as glucose dehydrogenase (or glucose oxidase) in the electrode proper increases the rate of charge transfer between the electrode and glucose. Thus, the enzyme effectively and selectively lowers the charge-transfer resistance. Both inorganic and organic catalysts have been used to selectively lower the charge-transfer resistance of selected substrates. *Selective membrane: Lower mass transfer resistances*: Let us assume that the analyte is an electrically neutral species, but the major interferant is charged. By placing an ion-exchange membrane with immobile charge of opposite polarity to that of the interferant in front of the electrode, the access of the charged interferant becomes blocked by the

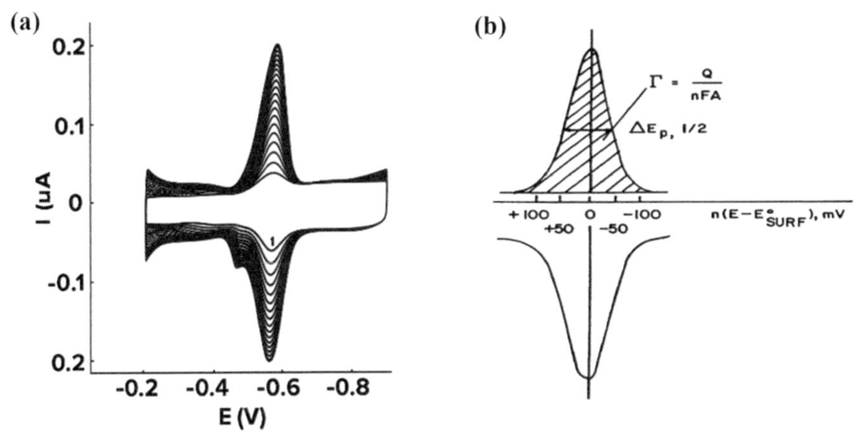

Fig. 7.16 **a** Surface confined process shows a cyclic voltammograms for 1×10^{-6} M riboflavin in a 1 mM sodium hydroxide solution. **b** Ideal cyclic voltammetric behavior for a surface layer on an electrode. The surface coverage, *T*, can be obtained from the area under the peak

electrostatic repulsion. These selectivity design strategies can be summarized in a statement that applies also to other life situations.

From the sensing point of view, the motivation for electrode modification has been to introduce additional flexibility in the design of, and additional control over, the electrochemical processes taking place at the electrodes, to facilitate a specific task, to lower the overall activation energy of the electron transfer for specific species and to increase or decrease the mass transport, or selectively accumulate the analyte.

Modified Electrodes

Electrostatic rejection of anions by negatively charged pores in the Nafion membrane. It is possible to selectively hinder the access of a species or a group of species to the surface of the electrode. This principle is illustrated on the example of a Pt micro-electrode coated with a thin layer of Nafion. Nafion is a sulfonated Teflon. It carries immobile negative charges. It is highly permeable to cations and to the neutral species. Such electrodes have been used successfully for in vivo monitoring of cationic neurotransmitters, such as dopamine, norepinephrine, and 5-hydroxytriptamine in the presence of anionic interferants, ascorbate, 3,4,3-dihydroxyphenylacetic acid, and 5-hydroxyindoleic acid, all of which are negatively charged at physiological pH. These anions are electrostatically excluded from the electrode surface by the negatively charged sulfonic groups present in the Nafion pores. It has been shown that the transport of the interfering anions to the electrode can be inhibited to 0.5–0.7% of the equivalent concentration of dopamine [2, 26, 28, 29] (Fig. 7.17).

Oxygen-Based Enzyme Electrodes

Type-1 design and operation of enzymatic amperometric sensors. "Dual" glucose sensor based on two oxygen electrodes, one (**a**) covered with active glucose oxidase and other (**b**) with deactivated glucose oxidase (Fig. 7.18). When glucose is present in solution, the enzymatic reaction takes place and the amount of oxygen reaching the surface of the "active" electrode is reduced by the amount consumed by the enzymatic reaction. Problems: Self-inhibition caused by the hydrogen peroxide. The second problem is encountered when the electrode is used in vivo limited by the availability of oxygen (Figs. 7.19 and 7.20).

Conductivity

In liquids, the conductivity varies according to the charge on the ion, the mobility of the ion, and the degree of dissociation of the ion. These all introduce complications. Conductivity is the *inverse of resistance*. It is a measure of the ease of passage of electric current through a solution. Conductance is related to the dimensions of a cell in a similar way to resistance. Conductometric and impedimetric methods are fundamentally non-selective methods. However, with the advent of modified surfaces for selectivity and much improved instrumentation these sensors have become more viable methods for designing biosensors. There are some very practical considerations that make the methods attractive for inexpensive, simple measurement methods—no reference electrodes are needed. The sensing units themselves can be fabricated at relatively low cost.

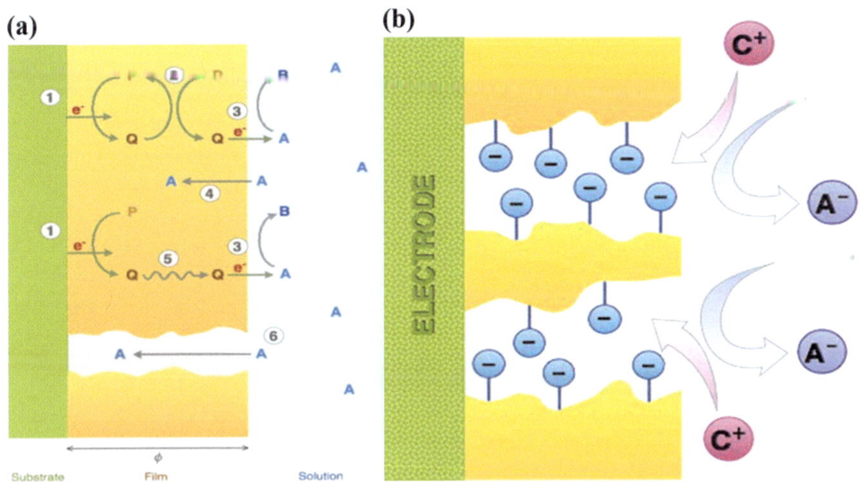

Fig. 7.17 **a** Schematic illustration of charge transport and ion diffusion processes in a redox-active polymer film. Electrons injected from the substrate reduce oxidized sites (P → Q), while counterions (A^-) from the solution migrate into the film to maintain electroneutrality. Mechanisms include electron hopping (steps 1–3), ionic diffusion (steps 4 and 6), and conformational or segmental motion (step 5), contributing to the overall charge transport. **b** Electrochemical doping/dedoping mechanism in a conducting polymer film. Upon application of potential, anions (A^-) and cations (C^+) from the electrolyte migrate into or out of the porous polymer film to compensate for injected charges, thereby modulating the film's conductivity and oxidation state

The measurement of conductance involves an alternating current, as in the classical conductance bridge. Varying the frequency of the alternating current can extend this. The quantity measured is then the *admittance* = 1/*impedance*. The admittance depends on simple conductance and the capacitance and inductance of the system. Using a frequency response analyzer, and then displaying in an Argand diagram can separate these components as imaginary components. Such diagrams are sometimes called *admittance (impedance) spectra*. This approach has not so far been used to any great extent in developing sensors and biosensors but is now receiving increased attention.

Direct Current (DC) Resistance: The electrical resistance of a circuit component or device is defined as the ratio of the voltage applied to the electric current which flows through it: The electrical resistance of a circuit component or device is defined as the ratio of the voltage applied to the electric current which flows through it: While Ohm's Law applies directly to resistors in DC or in AC circuits, the form of the current–voltage relationship in AC circuits in general is modified to the form:

$$V = I \times R \text{ for DC current}$$

$$V = I \times Z \text{ for AC current,}$$

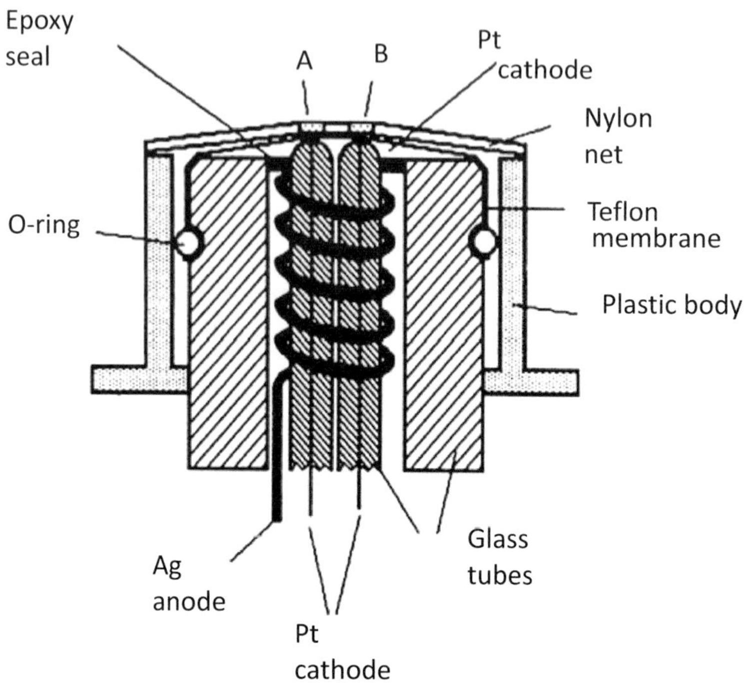

Fig. 7.18 Cross-sectional schematic of a Type-1 dual glucose sensor. The sensor features a coiled Ag/AgCl reference electrode embedded within a glass body. Region A represents the enzyme layer (e.g., glucose oxidase) immobilized on the sensor surface, while Region B denotes a selective membrane that permits glucose diffusion while excluding interfering species. This dual configuration allows for differential measurement of glucose levels, enhancing accuracy and specificity in complex biological environments

where I and V are the rms (root mean square current and voltage) or "effective" values.

The quantity Z is called impedance. For a pure resistor, $Z = R$. Because the phase affects the impedance and because the contributions of capacitors and inductors differ in phase from resistive components by 90°, a process like vector addition (phasors) is used to develop expressions for impedance.

For ordinary currents and frequencies, the behavior of a resistor is that of a dissipative element which converts electrical energy into heat. It is independent of the direction of current flow and independent of the frequency. Hence, we say that the AC impedance of a resistor is the same as its DC resistance (Fig. 7.21).

Complex AC Circuits

When capacitors or inductors are involved in an AC circuit, the current and voltage do not peak at the same time. The fraction of a period difference between the peaks expressed in degrees is said to be the phase difference. The phase difference is $\leq$ 90°. It is customary to use the angle by which the voltage leads the current. This

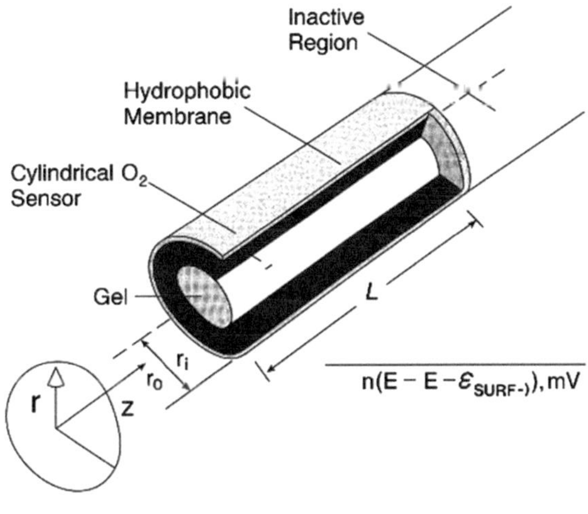

Fig. 7.19 Type-2 design and operation of enzymatic amperometric sensors. Oxygen, which is consumed by the primary glucose oxidation, is regenerated by the oxidation of hydrogen peroxide at the surface of the electrode. Because the operating potential is positive, oxygen is not consumed at the electrode. The reaction is not limited by the concentration of oxygen. Current is independent of the stirring rate, which means that the depletion layer is indeed confined to the interior of the gel layer. Two-dimensional, cylindrical glucose sensor. Glucose diffuses to the sensor only in direction z whereas oxygen can reach the electrode by diffusion through the walls of the cylinder of length L

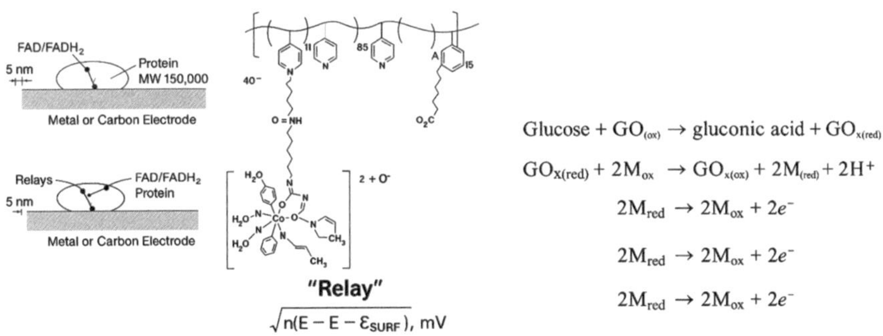

Fig. 7.20 Type-3 design and operation of enzymatic amperometric sensors—oxidation of glucose oxidase. The electrodes are modified with osmium redox polymers able to shuttle electrons from the prosthetic group of the enzyme toward the electrode. Considering the electron transfer pathways, glucose oxidase is re-oxidized in the last step by the redox polymer being able to oxidize multiple glucose molecules. The modified electrode is independent by oxygen allowing further miniaturization and implantation

leads to a positive phase for inductive circuits since current lags the voltage in an inductive circuit. The phase is negative for a capacitive circuit since the current leads the voltage. The useful mnemonic ELI the ICE man helps to remember the sign

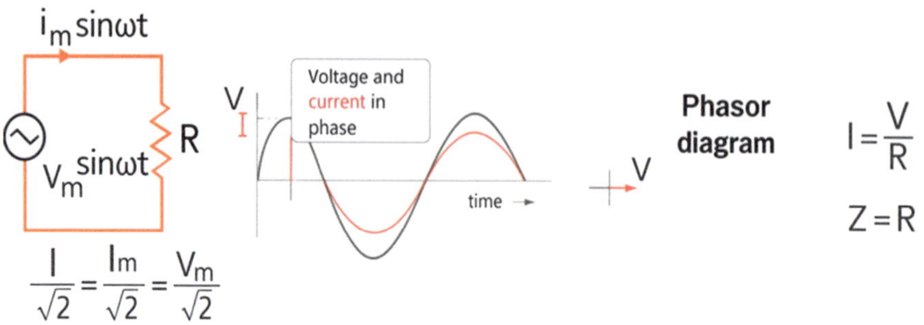

Fig. 7.21 Phasor diagram and time-domain representation for a purely resistive AC circuit. In a resistor (R), voltage (V) and current (I) are in phase, as shown by the overlapping sinusoidal waveforms and aligned phasors. The root mean square (RMS) values for voltage and current are $V_m/2\text{-}1$ and $I_m/2\text{-}1$

of the phase. The phase relation is often depicted graphically in a phasor diagram (Fig. 7.22).

It is sometimes helpful to treat the phase as if it defined a vector in a plane. The usual reference for zero phase is taken to be the positive x-axis and is associated with the resistor since voltage and current are in phase. The length of the phasor is proportional to the magnitude of the quantity represented, and its angle represents its phase relative to that of the current through the resistor. The phasor diagram for the RLC series circuit shows the main features (Fig. 7.23).

The voltage across a capacitor lags the current because the current must flow to build up the charge, and the voltage is proportional to that charge which is built up on the capacitor plates (Fig. 7.24).

The voltage across an inductor leads the current because the Lenz's law behavior resists the buildup of the current, and it takes a finite time for an imposed voltage to force the buildup of current to its maximum.

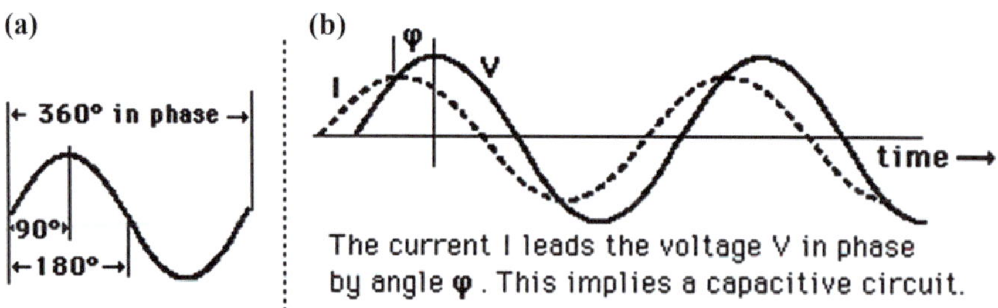

Fig. 7.22 Phasor diagram and time-domain response of a purely capacitive AC circuit. In a capacitor, the current leads the voltage by 90°, as shown by the phase shift in the sinusoidal waveforms and the corresponding phasor representation. The impedance of the capacitor is given by $Z = 1/j\omega C$, where ω is the angular frequency and C is the capacitance. This relationship characterizes the leading behavior of current in capacitive elements under AC excitation

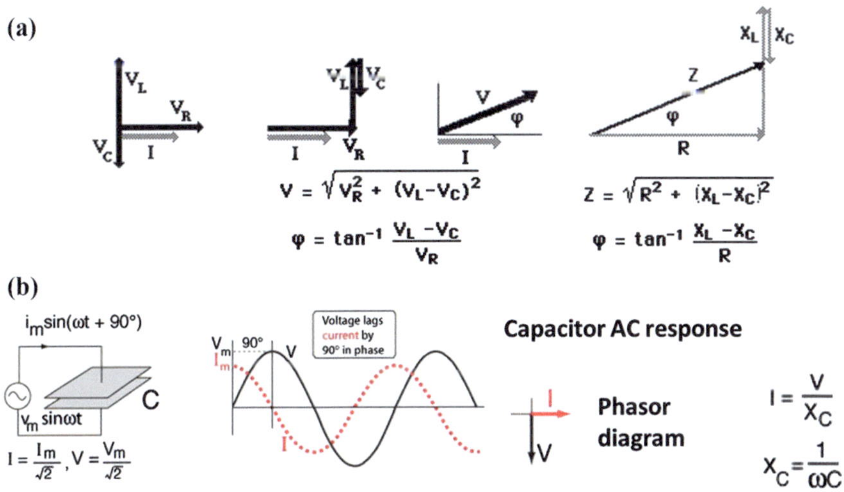

Fig. 7.23 **a** Phasor diagram for an ideal inductive AC circuit. In an inductor, the voltage leads the current by 90°, as shown by the sinusoidal waveforms and the vertical phasor shift. The inductive reactance is given by $X_L = \omega L$, where ω is the angular frequency and L is the inductance. **b** AC response of a capacitive circuit. The sinusoidal current leads the voltage by 90°, consistent with capacitive behavior. The phasor diagram confirms this phase shift with current and voltage vectors orthogonal to each other. The capacitive reactance is defined as $X_C = 1/\omega C$, and current is calculated using $I = V/X_C$

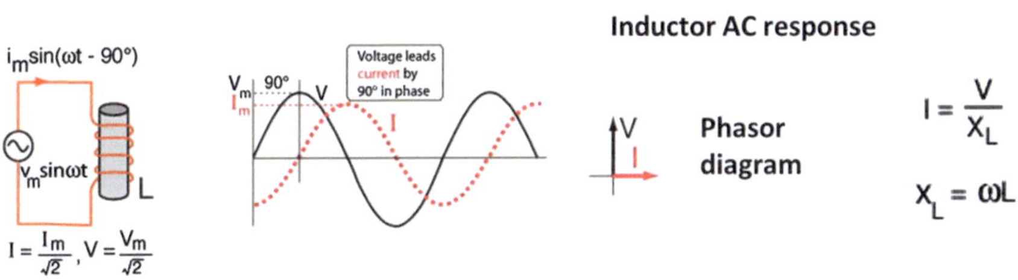

Fig. 7.24 Phasor diagram: inductor AC response illustrated through circuit diagram, sinusoidal waveforms, and phasor representation. In a purely inductive circuit, voltage leads current by 90°, as shown in the time-domain plot. The phasor diagram confirms this phase shift, with the voltage vector orthogonal and ahead of the current vector. The inductive reactance is defined as $X_L = \omega L$, and current is given by $I = V/X_L$, where ω is the angular frequency and L is the inductance

Impedance is graphed in the plane of complex numbers, which is divided into four quarters (quadrants) by the real and imaginary axes. Typical Nyquist plots in biosensing systems are shown in Fig. 7.25. (a) "Normal" electron transfer resistance, (b) small electron transfer resistance, (c) large electron transfer resistance.

Faradaic impedance spectra recorded upon sensing of foot-and-mouth-disease antibodies at a Au electrode functionalized with a self-assembled monolayer of the thiolated polypeptide corresponding to the foot-and-mouth-disease antigen: (a) bare

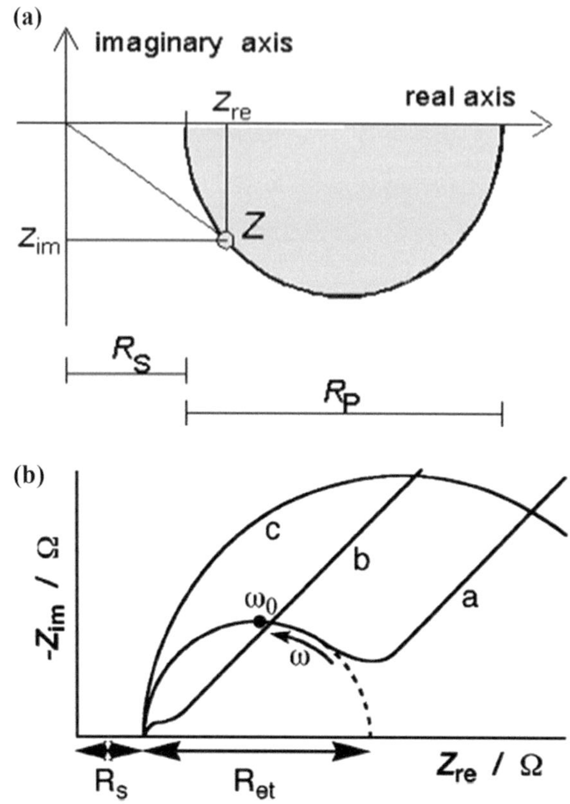

Fig. 7.25 **a** Nyquist plot representing the impedance spectrum of an electrochemical system. The real (Z_{re}) and imaginary (Z_{im}) components of the impedance are plotted on the horizontal and vertical axes, respectively. The semicircular arc corresponds to a charge-transfer resistance (R_P) and double-layer capacitance, while R_S indicates the solution resistance. The point Z represents the total impedance at a specific frequency. **b** Typical Nyquist plots for different electrochemical systems showing variations in impedance behavior. Curve (**a**) corresponds to a diffusion-limited process, curve (**b**) to a simple charge-transfer process, and curve (**c**) to a system with both charge transfer and capacitive behavior. R_S is the solution resistance, and R_{et} is the electron transfer resistance. The position of the semicircle's peak at frequency ω_0 provides insight into the characteristic time constant of the system

Au electrode. (b) Antigen-monolayer functionalized Au electrode prior to the reaction with the antibody. (c) Antigen-monolayer functionalized Au electrode after the reaction with the antibody. The data were recorded in the presence of $[Fe(CN)_6]^{3/4-}$, 2 mM [30] (Fig. 7.26).

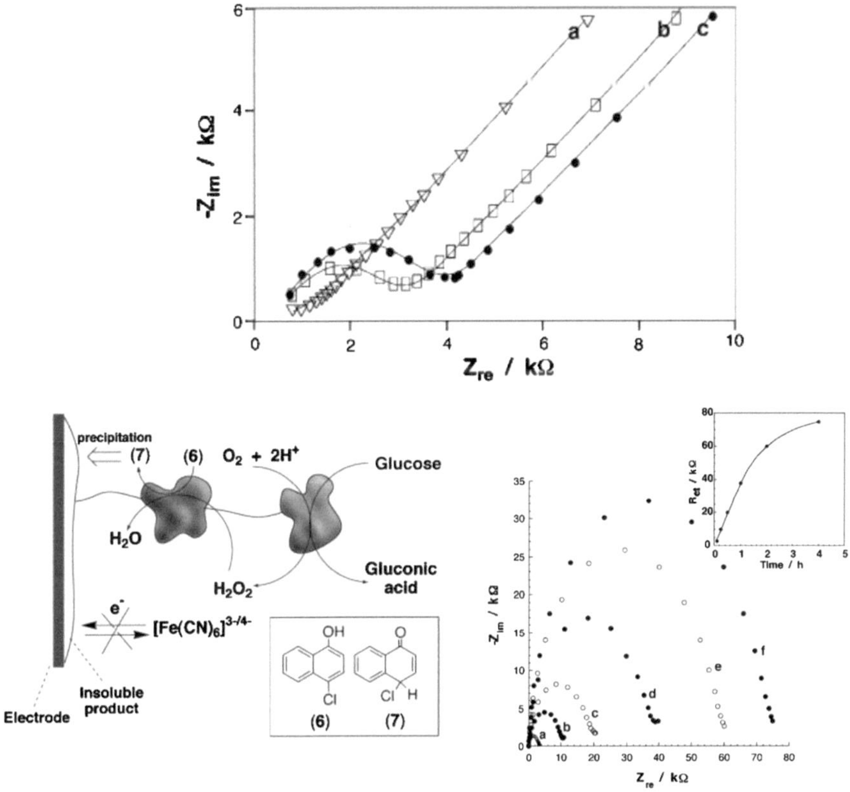

Fig. 7.26 (Top) Nyquist plots depicting the impedance spectra of an electrochemical system under varying conditions (curves a, b, and c). The plots show the imaginary component of impedance ($-Z_{im}$) versus the real component (Z_{re}). The increasing semicircle diameters from a to c indicate a progressive rise in charge-transfer resistance, possibly due to surface modification, fouling, or increasing analyte concentration. These changes reflect hindered electron transfer at the electrode interface. (Bottom left) Enzymatic glucose sensing mechanism involving glucose oxidase (GOx). Glucose is oxidized to gluconic acid while molecular oxygen is reduced to hydrogen peroxide (H_2O_2). The resulting H_2O_2 reacts with indicator compounds (e.g., compound 6: 4-chloro-1-naphthol), forming an insoluble colored product (compound 7: 4-chloro-1-naphthoquinone) that precipitates on the electrode surface. The ferricyanide/ferrocyanide redox couple ($[Fe(CN)_6]^{3-/4-}$) serves as an electron mediator in the detection process. (Bottom right) Nyquist plots (imaginary vs. real impedance) showing the evolution of interfacial impedance in a glucose biosensor over time (curves a–f). The increasing arc diameter indicates a rise in charge-transfer resistance, likely due to the accumulation of the insoluble product on the electrode surface. The inset plot depicts the time-dependent growth of resistance (R_{et}) over a 5-h period, confirming progressive surface blocking

7.2 Optical and Plasmonic Sensors

Interaction of electromagnetic radiation with matter occurs over a broad range of frequencies and usually in a highly specific way. Study and use of these interactions is in the domain of spectroscopy, which provides information ranging from the electronic structure of atoms to dynamics of polymeric chains. The interaction of this radiation with the sample is evaluated from the change of some optical parameter and is related to the concentration of the analyte. General arrangement of spectroscopic absorption/fluorescence measurement as shown in Fig. 7.27 includes *X*—light source, wavelength selector, *D*—detector, *R*—recorder. Intensity of initial (I_0), transmitted (I), or fluorescent (I_F) light is related to the concentration of the sample *C*.

The concept of dual character of light is particularly relevant to the discussion of optical sensors. Part 1: The basic quantized (corpuscular) aspects of light as they relate to optical sensors. Part 2: Optical waveguides and optical fibers which rely on wavelike (continuous) properties of light. Propagation of Electric (E) and Magnetic (H) vectors in light is shown in Fig. 7.28.

Light consists of two sinusoidally oscillating, perpendicularly oriented, electric (E) and magnetic (H) fields. They are in phase and move through space with velocity $c = 2.9979 \times 10^8$ m s^{-1} in vacuum. The frequency of oscillation v, wavelength A, and velocity C are related by

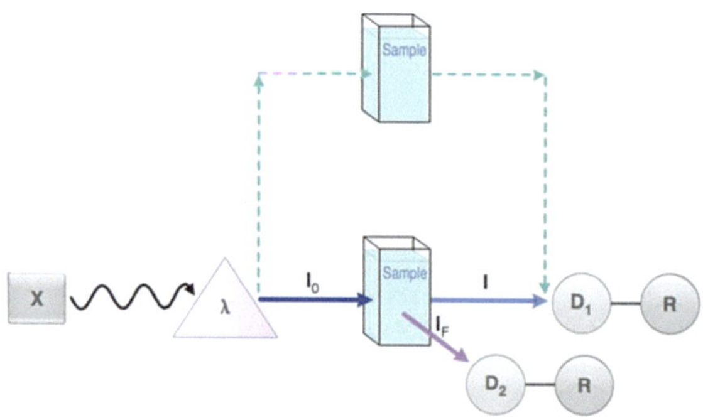

Fig. 7.27 General arrangement of spectroscopic absorption/fluorescence measurement

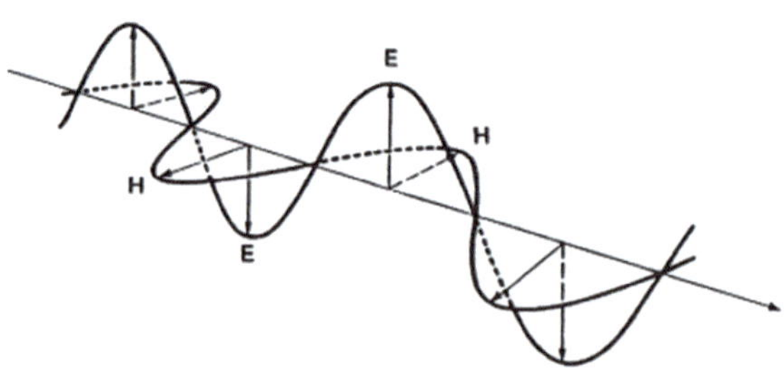

Fig. 7.28 Propagation of electric and magnetic vectors

$$c = \lambda \nu$$

When light interacts with matter several processes can take place, sometimes simultaneously. Light can be: (1) absorbed, leading to a decrease in the intensity of the primary beam, or *(2) transmitted* without attenuation. On reaching an interface light can be *(1) reflected* or *(2) refracted*—particularly important in guiding light through optical fibers and waveguides. If the direction and/or frequency of the light are changed upon interaction, we talk about *scattering. Fluorescence* and *phosphorescence* are special forms of scattering.

Light Sources Classification

Incandescent lamps (such as tungsten halogen lamps): These lamps emit over a broad spectral range (visible, IR) and are relatively inexpensive and are available in compact sizes. They are not, however, suited to modulation; they generate heat and have much shorter operational lifetimes than LEDs, for example. *Discharge lamps*: Deuterium lamps (200–300-nm range) and Xenon flash lamps (200–1000-nm range) require expensive power source and suffer from pulse-to-pulse variation in output intensity. *Lasers* (*non-semiconductor*): a very limited selection of lasers is available at specific wavelengths (Ar ion at 488 and 514 nm, HeNe at 633 nm, and HeCd at 325 and 442 nm).

Semiconductor sources (*LEDs, laser diodes*): Laser diodes (at 630–670 nm, 750–830 nm) are the most attractive option for sensors because of their low power consumption, high stability, long lifetime, robustness and compact size, and inexpensive, especially at visible and near-IR wavelengths.

Spectroscopic Principles: Most optical chemical sensors are based on a spectroscopic technique such as measurement of absorption, reflection, scattering, or fluorescence in the form of incident beam of intensity I_0 yielding: Transmitted (I_T), reflected (I_R), fluorescence (I_F) (Fig. 7.29).

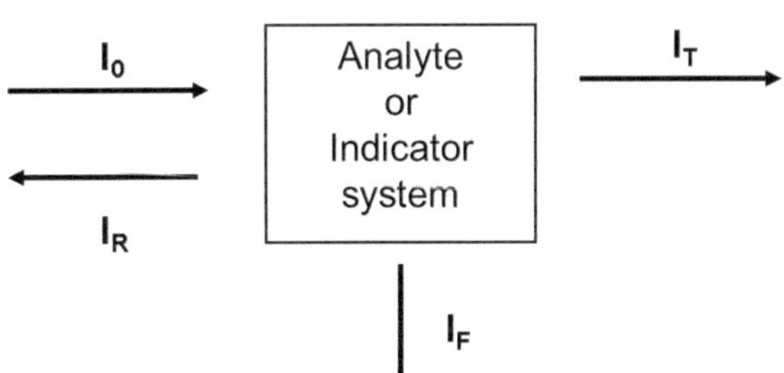

Fig. 7.29 Schematic representation of light interactions in an analyte or indicator system. The incident light intensity (I_0) interacts with the analyte, resulting in reflected (I_R), transmitted (I_T), and fluorescent (I_F) components. This diagram illustrates the fundamental processes of light absorption, reflection, transmission, and fluorescence relevant in optical sensing and spectroscopic detection

Absorption: Imagine that we have an electron in a lower energy state, and a photon comes along. If this photon has just the right amount of energy, it can be absorbed, causing the electron to make a transition to a higher-energy state. This will only occur though if the photon has an energy equal to the energy difference between the two levels involved in the transition—if not, the photon will just pass through. As the frequency of incoming light is scanned, the degree of absorption of the interacting light varies. The relationship between the relative absorption and the wavelength is called the *absorption spectrum* and is characteristic of the properties of the medium. Examination of the energy belonging to the different parts of the spectrum shows that these interactions will involve different properties of the molecule.

When a molecule absorbs a photon of an exact quantum of energy, it moves from its electronic ground state to its higher electronic excited state. Within each electronic state, the molecule can occupy different vibrational energy levels that are separated from each other by energy differences of less than 4 kJ mol^{-1}. The difference between two electronic levels is much larger than the difference between the individual vibrational levels. The ability of the molecule to absorb light is expressed as molar absorptivity ε_λ, which in analytical literature is also known as the extinction coefficient. When light passes through the sample with path length L and concentration of the absorbing species C, then absorption is characterized by the Beer–Lambert law:

$$I_T = I_0\, 10^{-eLC},$$

where e is the molar absorptivity, or extinction ($mol^{-1}\ cm^{-1}$) of the absorbing species at λ_{max}, L is the absorption path length (cm), and C is the concentration (mol/L) of the absorbing species. The absorbance of the sample or indicator system is shown in Fig. 7.30.

The most important practical consequence of this equation is that when I_T and I_0 are known, the calculated absorbance A is directly proportional to the concentration C, thereby yielding a linear calibration. Some spectroscopists prefer to use the equivalent

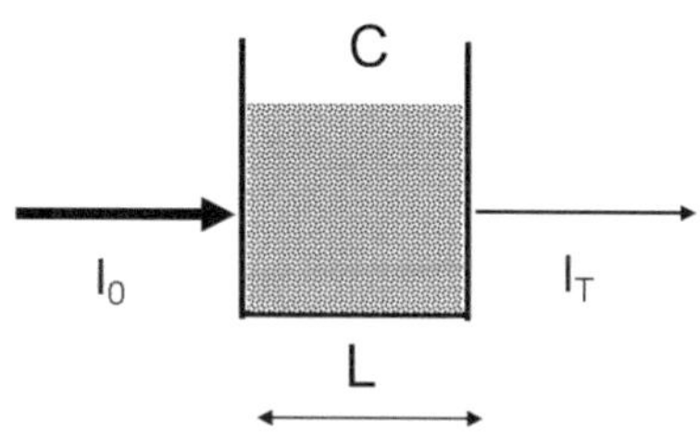

Fig. 7.30 Schematic diagram of a typical absorbance-based optical detection setup. A light beam with initial intensity I_0 passes through a cuvette (C) containing a solution of analyte with concentration C and path length L. The transmitted light intensity I_T is measured on the opposite side. This arrangement follows the Beer–Lambert law, which relates absorbance to analyte concentration, path length, and molar absorptivity

expression, $I_T = I_0 \exp(-\alpha L)$ where α is the absorption coefficient of the absorbing material.

Molar absorptivity is a *function of wavelength* and is an *additive property*. This means that for samples containing multiple absorbing species within the same optical path and at the same wavelength, the absorbance A is given by

$$\log I_0/I = A = L(e_1C_1 + e_2C_2 + \cdots + e_nC_n),$$

where the e_C quantities are the contributions of the individual species to the total attenuation. The shapes of the absorption spectra of different species vary (i.e., the e quantities are different functions of λ), so equation serves as a powerful tool in the quantitative spectrophotometry of complex mixtures and the possibility to perform multi-wavelength measurements. There are practical limitations to the applicability of the Beer–Lambert (*BL*) law, in particular, the linear dependence of absorbance on concentration: (1) The bandwidth $\Delta\lambda$ of the incident beam should be very narrow, ideally approximating monochromatic radiation. (2) Deviations from perfect *BL* behavior are also observed in highly absorbing or highly scattering media. Both effects yield a very limited linear range for the absorbance–concentration relationship.

Reflectance: The reflectance $R\,(= I_R/I_o)$ is related nonlinearly to C via the Kubelka–Munk function F_{KM}, which is given by

$$F_{KM}(R) = \frac{(1-R)^2}{2R} = \frac{\varepsilon C}{S},$$

where S is a scattering coefficient. Although diffuse reflectance spectroscopy is a well-established analytical technique with particular application in near-IR analysis of powders and other opaque materials, the use of this technique in sensors has been very limited.

Fluorescence: When a molecule absorbs a photon possessing an exact amount of energy, it moves from its electronic ground state to its first excited state. If the molecule can remain in the lowest vibrational level of the excited state long enough, it can make the transition to the ground state either by a strong collision with another molecule or radiatively. In the latter case, the energy loss occurs in a quantized fashion and the emitted light again has a characteristic *fluorescent* frequency. In general, molecules in gas and in liquid phase do not have a fixed orientation with respect to the exciting light. Therefore, the emission of fluorescence occurs in all directions. The emitting molecule is called ***fluorophore***. Energy dissipation by strong collisions is clearly a competing process, which limits the efficiency of fluorescence (Fig. 7.31).

Different molecules possess different ability to dissipate energy from the fluor, partly because of their concentration but also because of their stereo specific fit to the excited molecule.

Molecules which can dissipate energy from an excited molecule by a thermal process even at low concentrations are called **quenchers**. The fluorescence intensity

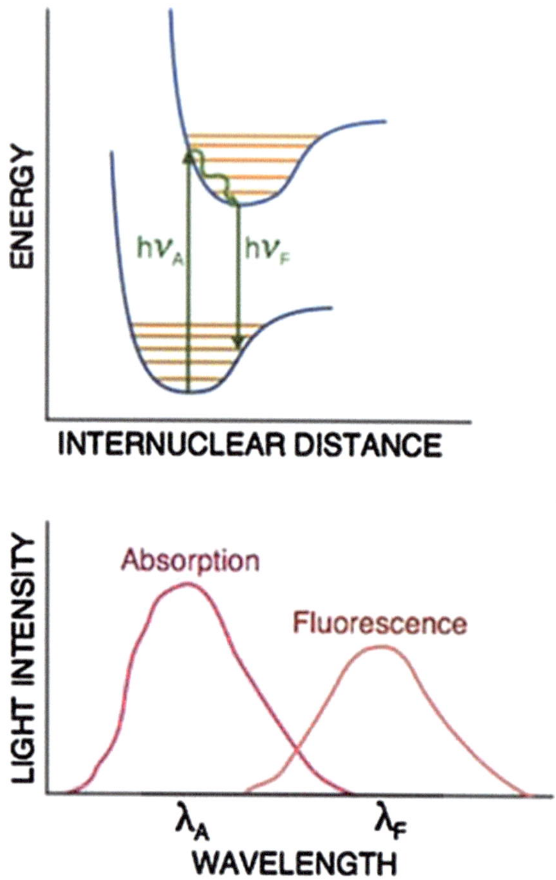

Fig. 7.31 Fluorescence: a diagram illustrating the mechanism of fluorescence-photon absorption causes an excitation followed by non-radiative and radiative relaxations

I_F depends on the quantum efficiency ζ given by

$$\zeta = (\text{photons emitted})/(\text{photons absorbed}).$$

It also depends on the concentration of the fluor C, on its molar absorptivity ε, and on the path length L.

Usually only a fraction of the emitted light is captured by the detector; hence, a collection factor γ is also introduced. The final expression is therefore

$$I_F = \gamma\zeta\varepsilon CL.$$

Fluorescence is the radiative deexcitation of a molecule flowing absorption of a photon. The emitted photon is of lower energy than the absorbed photon. The fluorescence emission peak of a species is at a higher wavelength than the absorption peak. The wavelength separation between these peaks is called the *Stokes shift*.

Fluorescence Quenching: While the excited molecule is undergoing radiation-less decay, it may also have time to react with another species, thus transferring its energy to the new chemical species and returning to the ground state with a diminished fluorescent output.

This process is known as quenching and provides another powerful way of using fluorescent behavior:

Fluorescence k_r

$$M* \rightarrow M + h\theta.$$

Quenching k_q

$$M * + Q \rightarrow M + Q * .$$

The fluorescent intensity is decreased to I_r from the initial absorption intensity I_0. The quenching effect is described by the Stern–Volmer relationship, as follows:

$$\frac{I_{F0}}{I_F} = 1 + K_{SV}[Q],$$

where I_{F0} and I_F are respectively the fluorescence signals in the absence and presence of quencher, $[Q]$ is the quencher concentration, and K_{SV} is the Stern–Volmer quenching constant. This constant is given by:

$$K_{SV} = kt_0,$$

where k is the diffusion-dependent bimolecular quenching coefficient and t_0 is the fluorescence decay time of the species in the absence of the quencher.

Phosphorescence: The transition to the first excited (singlet) state on absorption of the quantum of light is the same. If there is a triplet state (with two unpaired electrons) lying energetically close to or below the first excited state, a non-radiative singlet–triplet transition can take place. The radiative decay from the triplet state to the ground state, analogous to the fluorescent emission, then takes place. Because there are multiple rearrangements that must occur during this process, phosphorescence has much longer lifetime (up to milliseconds) than fluorescence (Fig. 7.32).

Luminescence: Chemiluminescence: The excited state is created by a chemical reaction, which then emits light according to the following scheme.

$$A + B \Rightarrow X* \Rightarrow X + \eta\nu.$$

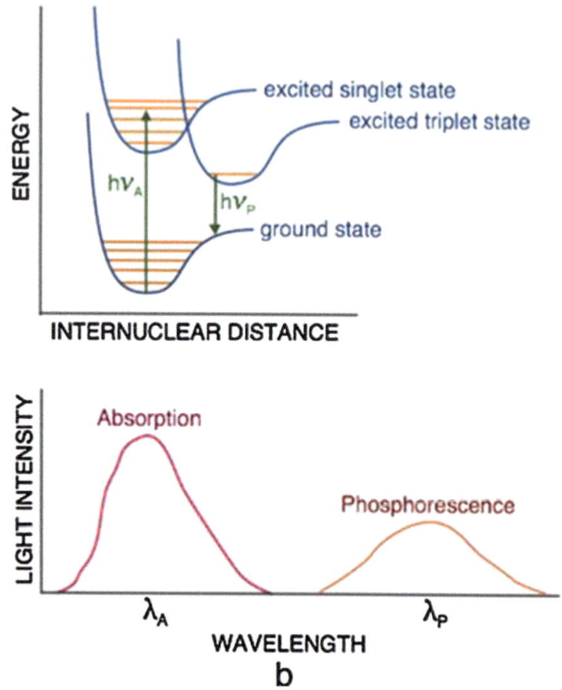

Fig. 7.32 Phosphorescence: a diagram illustrating an intersystem crossing into a triplet state, which then relaxes slowly to the ground state, emitting photons in the process

Electroluminescence: There are also several electrochemical reactions in which the electroluminescence occurs because of chemical reactions following the initial electron transfer. Bioluminescence: If biomolecules are involved, the emission is called bioluminescence. Mostly used as bioassays and chemical assays.

Scattering: In addition to being absorbed and emitted by atoms and molecules, photons may also be scattered (approx. 1 in 10^7 in a transparent medium). This is not due to defects or dust but a molecular effect which provides another way to study energy levels.

This scattering may be: **Elastic** and leave the molecule in the same state (Rayleigh scattering). **Inelastic** and leave the molecule in a different quantum state (Raman scattering).

When the sample is irradiated with light of a higher frequency (higher energy) than would correspond to any electronic transition, most of the light passes through the sample without attenuation. Only a small fraction is scattered, giving rise to very faint lines corresponding to the transitions within vibrational levels. These lines belong to the **Raman spectrum** and are observed at a right angle to the incident beam. These transitions are "not allowed," in the sense of the Bohr model of atom. However, if we realize that, in quantum mechanics, "allowed" means "having a higher probability" and, conversely, "not allowed" means "having a lower probability," then the existence of these transitions can be rationalized. Interaction of light with the molecule in which only the direction of propagation is changed is called **Rayleigh elastic scattering** as shown in Fig. 7.33. If the molecule absorbs a fraction of energy, weak Raman lines shifted toward lower frequencies (longer wavelengths) appear. They are called Stokes lines. Symmetrically with respect to the Rayleigh line lie the somewhat less intense anti-Stokes lines which correspond to the transfer of energy from the molecule to the incident light. Closer to the central line are weak Brillouin lines which are caused by the interaction of the light with the phonons in the condensed (i.e., liquid or solid) medium.

Elastic Scattering/Rayleigh: Rayleigh scattering refers to the scattering of light off of the molecules of the air and can be extended to scattering from particles up to about a tenth of the wavelength of the light. It is Rayleigh scattering off the molecules of the air which gives us the blue sky. Lord Rayleigh calculated the scattered intensity from dipole scatterers much smaller than the wavelength to be. Rayleigh scattering can be considered to be elastic scattering since the photon energies of the scattered photons is not changed.

Inelastic Scattering/Raman: These lines are called Raman lines (after the Indian physicist Chandrasekhar Venkata Raman). Raman scattering can occur with a change in vibrational, rotational energy of a molecule. Scattering in which the scattered photons have either a higher or lower photon energy is called Raman scattering. Usually, this kind of scattering involves. *(1) Exciting some vibrational mode of the*

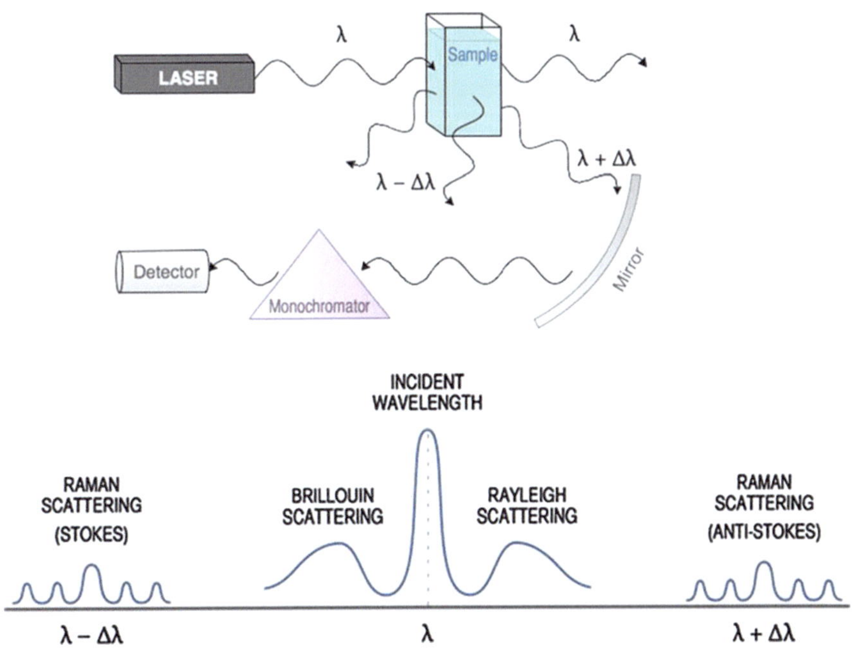

Fig. 7.33 Schematics of scattering 1 in 10^7 in a transparent medium occurring not due to defects or dust but a molecular effect to study energy levels

molecules, giving a lower scattered photon energy, (2) scattering off an excited vibrational state of a molecule which adds its vibrational energy to the incident photon. These weaker lines are observed at wavelengths which are shifted compared to the wavelength of the light source. A photon with some energy strikes the sample and the system is exited to a virtual energy level.

The majority of the photons incident on the system are scattered at the incident frequency; this is known as Rayleigh scattering or elastic scattering. A small portion of the incident light is scattered inelastically (at different energy) to upper vibrational levels of the ground electronic state. This is known as Raman Scattering. The difference in energy between the incident light and the Raman scattered corresponds to vibrational energy spacing in the molecule. The vibrational energy of the molecule is increased after the collision; the energy of the scattered photons is decreased for the same amount and, therefore, can be detected at longer wavelengths (red arrow). Stokes lines are those in which the photon has lost energy to the molecule. The respective spectral lines are called Stokes-shifted. If the vibrational energy of the molecule is decreased after the collision, the energy of the scattered photons is increased for the same amount and, therefore, can be detected at shorter wavelengths (blue arrow). Anti-Stokes lines are those in which the photon has gained energy from the molecule. The respective spectral lines are called anti-Stokes shifting. This is only possible if the molecule is at an exciting vibrational state before the collision (Fig. 7.34).

Raman Spectroscopy: The difference in energy between the incident photon and the Raman scattered photon is equal to the energy of a vibration of the scattering molecule. A plot of intensity of scattered light versus energy difference is a Raman spectrum. Since molecular energy levels are quantized this produces discrete lines

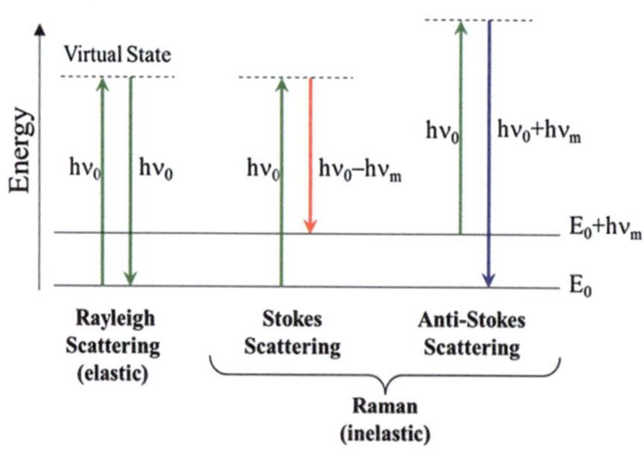

Fig. 7.34 Energy-level diagram illustrating the principles of light scattering in spectroscopy. Rayleigh scattering (elastic) involves no change in energy, with the scattered photon having the same frequency ($h\nu_0$) as the incident light. Stokes scattering (a form of Raman scattering) occurs when the scattered photon has lower energy ($h\nu_0 - h\nu_m$) due to energy transfer to the molecule. Anti-Stokes scattering involves scattering from an excited vibrational state, resulting in a higher-energy photon ($h\nu_0 + h\nu_m$). Raman scattering is thus inelastic and provides molecular vibrational information

from which we can gain info on the molecule itself. Like identification of minerals and organic substances, their chemical formulas and the arrangements of the atoms within them. Thus, whether the mineral was a carbonate, sulfate, phosphate, silicate, oxide, sulfide, hydroxide can be determined. Figure 7.35 shows an image of portable Raman spectrometer and representation of excitation source and scattered beam are collected by the optic fiber of the probe.

Surface-Enhanced Raman Scattering SERS: SERS was accidentally discovered while people tried to do Raman on the electrode in 1974. The original idea was to generate a high surface area on the rough metal. In 1977, the rough sliver electrode produce Raman spectrum that is a million-fold more intense than was expected. This enormously strong signal debuted surface enhanced Raman scattering (SERS). SERS can be used to detect single molecules. The rather weak Raman effect can be greatly strengthened (by a factor of up to 14 orders of magnitude) if the molecules are attached to nm-sized metal structures. (The way in which the enhancement occurs is still not known for sure.) SERS is a powerful phenomenon involving interaction of light, molecules, and metal nanostructures to enhance Raman signals that can resolve the structures down to single molecule level.

The enhancement occurs due to the resonant interaction between optical fields and surface plasmons in the metal. When incident light excites surface plasmons, they radiate a dipolar field, which together in turn leads to redistribution of electric field around the metal clusters. This effect can cause enhanced excitation intensity from the molecule adsorbed on the metal surface and shift in the frequency between incident and excited laser light. The advantages of this method are that it is fast, it can supply some structural information about the molecules, and it does not bleach the molecules. Single-molecule detection is of great practical interest in chemistry, biology, and medicine, and pollution monitoring; examples include DNA sequencing and the tracing of biomedically interesting molecules.

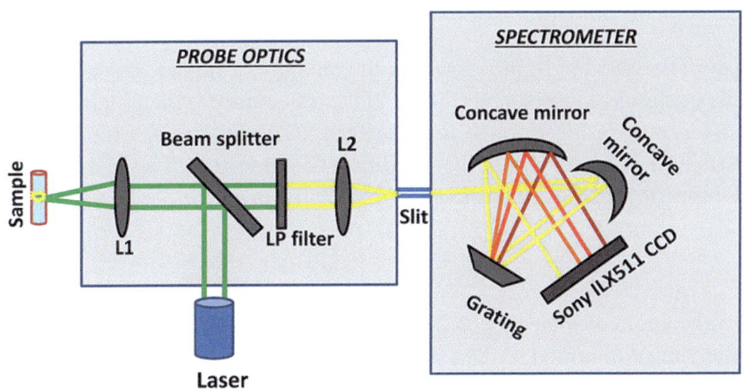

Fig. 7.35 Schematics of portable Raman spectrometer and excitation source and scattered beam are collected by the optic fiber of the probe

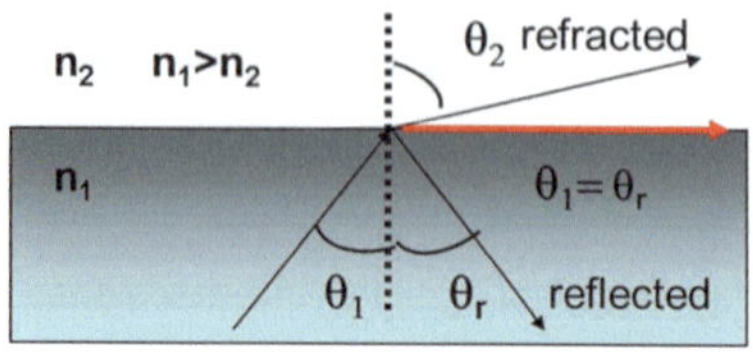

Fig. 7.36 Diagram illustrating reflection and refraction of light at an interface between two media with different refractive indices ($n_1 > n_2$). The incident angle (θ_1) equals the reflected angle (θ_r) according to the law of reflection. The refracted ray bends away from the normal ($\theta_2 > \theta_1$) as it passes from the denser medium (n_1) to the less dense medium (n_2), consistent with Snell's law: $n_1 \sin \theta_1 = n_2 \sin \theta_2$

Wave Properties and Optic Fibers

A typical optical sensor consists of modules found in conventional spectroscopy, source, monochromator, cuvette, and detector. The big difference is that light travels between these modules through optical fibers or waveguides. As it travels, the beam of light interacts with interfaces between media of different optical densities and therefore of different refractive indices. The laws that govern its behavior at these boundaries are the most important aspects of the operation of optical sensors. Optical density: For EM wave, the speed of wave depends on the OD of the material. OD is indicated by refractive index. Denoted by "*n*." RI values are represented relative to the speed of light. RI for vacuum is 1.0, air = 1 and for water = 1.33 (high OD). When light arrives through an optically dense medium of refractive index n_1 at its boundary with an optically rarer medium of refractive index n_2 (i.e., $n_1 > n_2$), it can be either ***Reflected*** back to phase 1 or ***Refracted*** (transmitted through) to phase 2, which of these two conditions applies is determined by the ***angle of incidence*** at the interface as shown in Fig. 7.36.

As the angle of incidence increases, the angle of refraction increases. At certain angle, the refracted ray follows a path exactly along the surface of the dense medium. When the angle of refraction is at 90, the incident angle is called the "critical angle."

Snell Law: The speed of light is lower in the medium of higher optical density. The Maxwell equations require that the phase of the transmitted light q_2 and its frequency be preserved on passing through the interface. The angle of the transmitted light must change in order to satisfy the condition of phase matching. The geometrical relationships for refracted light are given by Snell's law:

$$n_1 \sin q_1 = n_2 \sin q_2.$$

When the angle of incidence is increased beyond certain value, the condition of phase matching can no longer be satisfied and all the light is transmitted along the interface (i.e., $q_2 = 90°$). The angle at which this happens is called the *critical angle* q_c for which reduces Snell's equation to $\sin q_c = n_2/n_1$. For the air–water interface, q_c is 43.75°. For values $q_1 > q_c$, all the light is reflected to the optically dense phase. This is called the condition of *total internal reflection* (TIR).

Evanescent Wave: When light propagates in an optical fiber or waveguide, a fraction of the radiation extends a short distance from the guiding region into the medium of lower refractive index that surrounds it. In order to propagate an electromagnetic wave along the interface or to reflect it totally, there must be a zero net flux of energy into the optically rare medium (n_2).

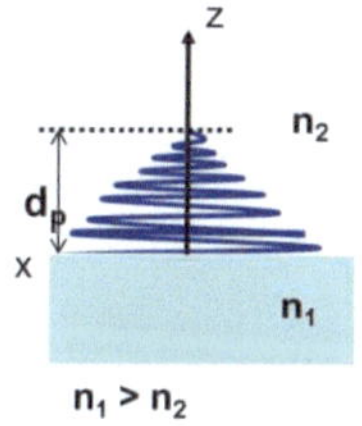

However, there is a finite decaying electric field across the interface which extends some distance x into the rarer medium. The intensity of this field is given by

$$E_2 = E_0 \exp\left[\frac{-\mathrm{i}2\pi n_2}{\lambda}(x \sin \Theta_2 + z \cos \Theta_2)\right].$$

When $\Theta_1 = \Theta_c$, and $\Theta_2 = 90°$, this equation reduces to:

$$E_2 = E_0 \exp\left(\frac{-\mathrm{i}2\pi n_2}{\lambda}x\right).$$

The reflected beam combines with the incident beam to form a standing electromagnetic wave at the interface. The electric field, which penetrates the optically rare medium, is called the ***evanescent field***. The depth of its penetration d_p is then defined as the distance at which the initial intensity decays to its $1/e$ value; d_p is defined as:

$$d_p = \frac{\lambda}{2\pi\left(n_1^2 \sin^2 \Theta_1 - n_2^2\right)^{1/2}}.$$

The depth of penetration plays a critical role in the definition of the "optical volume" that is being sampled. It depends on several parameters, above all on the wavelength of the incident beam, on the angle of incidence, and on the refractive indices of the two media. This consideration is important in sensing applications in which the interaction of a chemical species with this field gives rise to an analytical signal. For common materials and for visible light, d_p is typically between 100 and 200 nm. The evanescent field, which decays exponentially with distance from the waveguide interface, defines a short-range sensing volume within which the evanescent energy may interact with molecular species. Advantages include: no coupling optics are required in the sensor region because the interrogating light remains guided; considerable miniaturization is possible; the technique can provide enhanced sensitivity over conventional bulk-optic approaches; EW spectroscopy is suitable to perform accurate absorption measurements on highly absorbing or highly scattering media.

The optical core of refractive index n_1 (a typical material is silica, which has $n_1 = 1.6$) is surrounded by an optical insulator, called cladding, which has lower refractive index n_2 (a typical value is 1.5). The outside coating on the fiber is there for protection and has no special optical function. In order for the light to be coupled, it must enter the end of the fiber within a certain cone whose angle is related to the critical angle for reflection at the core-cladding interface. Likewise, light leaving the end of the fiber spreads only over the same solid angle. For a rational design of an optical sensor, it is important to know the **shape of this cone**. For the light to be coupled, it must enter the end of the fiber within a certain cone whose angle is related to the critical angle for reflection at the core-cladding interface. Likewise, light leaving the end of the fiber spreads only over the same solid angle. For the rational design of an optical sensor, it is important to know the shape of this cone. The *numerical aperture* NA of the fiber defines the cone of acceptance in terms of refractive indices alone: Optical fibers and waveguides can transmit light over large distances and with minimal loss of intensity. An evanescent field which exists at the surface of an optical fiber may be used for sensor application (Fig. 7.37).

The phenomenon of light reflection has two important implications for chemical sensing: Transmission of light through waveguides and fibers. An evanescent field on the outside of the waveguide, which represents a sampling space in which the chemical interactions can take place. When a single beam of light is introduced into a waveguide surrounded by a homogeneous medium of lower refractive index, light bounces from the opposite, interfaces and propagates in a zigzag fashion through the planar waveguide or as a spiral through the cylindrical optical fiber. When a single beam of light is introduced into a waveguide surrounded by a homogeneous medium of lower refractive index, light bounces from the opposite, interfaces and

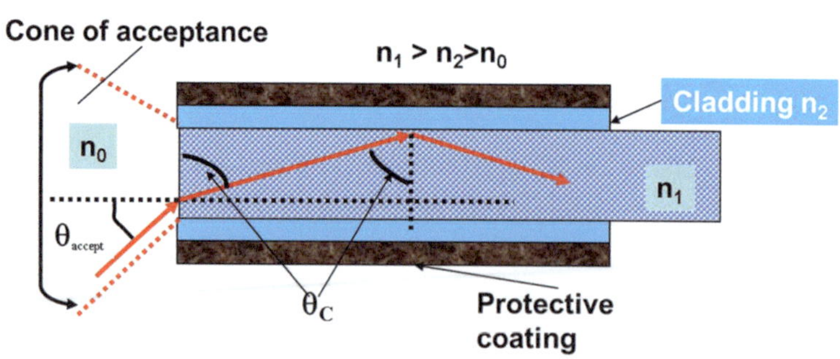

Fig. 7.37 Schematic diagram of light propagation in an optical fiber illustrating the principle of total internal reflection. Light entering the fiber within the cone of acceptance (defined by the acceptance angle θ accept) is guided through the core (refractive index n_1) by total internal reflection at the core-cladding interface ($n_1 > n_2$). The cladding (n_2) is surrounded by a protective coating, and light from an external medium (n_0) must enter at angles smaller than the critical angle (θ_C) to be efficiently transmitted through the fiber

propagates in a zigzag fashion through the planar waveguide or as a spiral through the cylindrical optical fiber. At each point of reflection, there is an evanescent field. If there is no interaction within this field and/or within the waveguide, the light will pass without attenuation. If decrease the thickness and "squeeze" the light, the points of reflection move closer to each other. As they begin to overlap, the reflected beams begin to interfere. This produces another set of conditions, which must be satisfied from the viewpoints of both the single reflection and the constructive interference. As a result of this additional constraint, a finite number of reflection angles exist for each wavelength at which the conditions allow propagation by **multiple total reflection**. These constructive, allowed patterns are called *guided modes* of the waveguide. The thinner the waveguide and the smaller the difference between the refractive indices, clearly the fewer the modes. When the waveguide operates in a multiple reflection mode, light no longer exists as an individual beam but rather appears as a continuous streak. The second consequence of increasing the number of multiple reflections is the appearance of the continuous evanescent field, which results from overlap of the individual evanescent points. This may serve in sensing applications as a large "active area" in which the sensing interaction can take place (Fig. 7.38).

For a non-coherent source, the light propagated by a fiber again appears as a continuum of intensity. The number of modes M in a fiber of radius r (both r and λ are in μm) is then given as:

$$M = \frac{(2\pi r)^2 \left(n_{\text{co}}^2 - n_{\text{cl}}^2\right)}{2\lambda^2},$$

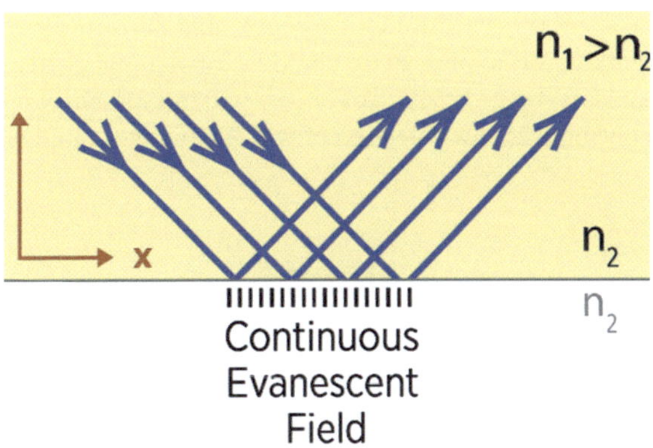

Fig. 7.38 Illustration of total internal reflection and generation of an evanescent field at an interface between two media with refractive indices $n_1 > n_2$. When light propagates in the higher-index medium (n_1) and undergoes total internal reflection at the boundary with the lower-index medium (n_2), a non-propagating evanescent field is generated along the interface. This evanescent wave penetrates a short distance into the lower-index medium and is fundamental to techniques like evanescent wave spectroscopy and optical fiber sensing

where n_{co} and n_{cl} are the refractive indices of the fiber core and cladding, respectively. A typical fiber of $d < 10$ μm supports one mode (a single-mode fiber), while thousands of modes can be accommodated in fibers which are approximately 100 μm in diameter. M is characterized by the V-number, which is a practical figure of merit for the fiber as shown below

$$V = \frac{2\pi r}{\lambda}\sqrt{n_{co}^2 - n_{cl}^2}.$$

Surface Plasmon Resonance

When the interface between a thin (~ 50 nm) film of free electron metal (e.g., Au, Ag, or Al) and a glass prism is irradiated with a TM polarized coherent light, correlated motion of free electrons in the metal leads to the formation of oscillating charge at the metal/glass interface A corresponding oscillating charge also appears at the other interface between metal and the sample, which has dielectric constant ε_d. Collectively, the two oscillating vectors are known as **surface plasmon** which moves along the surface, but does not radiate energy (Fig. 7.39).

Its intensity decays exponentially to the dielectric (sample) in the direction perpendicular to the interface. In that respect, it is like the **evanescent field** [31–35].

Applications on affinity-based sensors. Optical biosensors have been particularly attractive and widely investigated for detecting target analytes due to their ability toward minute changes in the mass and thickness on the surface of the transducer. These sensors show high sensitivity, fast response and adaptability to various conditions. Most commonly employed optical techniques are fluorescence/ phosphorescence spectrometry, bioluminescence, and surface plasmon resonance. Based on their working principles, these could be labeled, label-free, "turn-on" and "turn-off" mechanisms. In the below section, we will discuss novel optical biosensor strategies based on aptamers for detection of various analytes ranging from proteins,

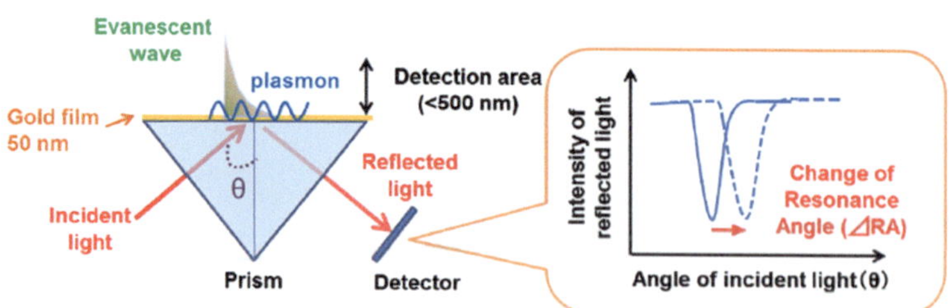

Fig. 7.39 Schematic of surface plasmon resonance (SPR) sensing using the Kretschmann configuration. Incident light passes through a prism and strikes a thin gold film (~ 50 nm), generating an evanescent wave that excites surface plasmons at the metal–dielectric interface. The reflected light intensity is monitored by a detector as a function of the incident angle (θ). Resonance occurs when the incident light couples with surface plasmons, causing a sharp dip in reflected intensity. Shifts in the resonance angle (ΔRA) indicate changes in the refractive index near the sensor surface, enabling label-free detection of molecular interactions

drugs, ions, and small molecules. Most of the optical sensors are constructed based on thrombin and OTA as the model target.

Additionally, a target-induced structure switching aptamer strategy was developed for detection of ochratoxin A. OTA aptamer labeled with fluorescein hybridizes with complementary DNA and quencher moiety giving to highly efficient quenching; however, in the presence of OTA conformational changes occur and a strong concentration-dependent fluorescent signal is observed. A wide linear detection range (from 1 to 100 ng/mL) was achieved with a detection limit down to 0.8 ng/mL which also exhibited high selectivity among other mycotoxins. Novel turn-off luminescence resonance energy transfer (LRET) aptamer sensor was designed to detect thrombin. This strategy is based on the overlap between the emission spectrum of the $NaYF_4:Yb^{3+}$, Tm^{3+} nanophosphors, and the absorption spectrum of the gold nanorods. Negatively charged PAA@UCNPs (polyacrylic acid-upconverting nanophosphors) with thrombin aptamer interact with the positively charged gold nanorods via electrostatic interactions to induce quenching of the luminescence. In the presence of thrombin, the luminescence is recovered due to conformational changes that displace PAA@UCNPs. The linear range was observed from 3.75 to 112.50 nmol/L [36].

Another study on exonuclease III (Exo-III)-aided amplification and graphene oxide (GO) platform for fluorescence quenching, a novel, turn-on fluorescent aptasensor for lysozyme (Lys) protein was constructed. This novel strategy contains a system with hairpin probe (HP) and a signal probe (SP) labeled with carboxy fluorescein (FAM) with lysozyme aptamer. Lysozyme binds with this system triggering Exo-III-aided amplification and liberation of free fluorophore. Presence of GO led to adsorb the intact SPs and leading to fluorescence quenching which was detected with linear range from 0.125 to 1 mg/mL concentrations of lysozyme [37].

A similar approach using exonuclease III was employed as a signal amplifying system for construction of fluorescent based aptasensor for adenosine detection. The adenosine aptamers were hybridized with the complementary DNA (cDNA), and the Exo-III could not cleave the single-strand signal probes labeled with carboxyl fluorescein (FAM) at its 5′ ends when adenosine was absent. Later when GO is added, the fluorophore was quenched. Contrary to that when adenosine was present aptamer association with the targets, formation of DNA duplex, and finally Exo-III-induced cleavage liberation of fluorophore occurred giving remarkable fluorescent increase. The limit of detection was found to be 1 nM [38]. A novel label-free calorimetric aptasensor was developed based on nicking enzyme assisted signal amplification and DNAzyme amplification for highly sensitive detection of thrombin was reported specifically with a detection limit as low as 1.5 pM, which is at least four orders of magnitude lower over the unamplified colorimetric assay. For the first time, a label-free electrochemiluminescence (ECL) aptasensor with a strategy based on novel assembly of oligonucleotide and luminol functionalized gold nanoparticles (luminol-AuNPs was constructed [39]. An Au/AuNPs/ds-DNA/luminol-AuNPs/TBA multilayer architecture acts as sensing interface and undergoes electro-oxidation reaction of luminol. In the presence of thrombin, TBA captures it onto the electrode surface which produces a barrier in electron transfer and decreasing the luminescence signal which is correlated to the concentration of thrombin. Thus, a wider linearity for the detection of thrombin in the range of 0.005–50 nM with a detection limit of 1.7 pM was obtained. Potassium ion K(+) was detected using based on the horseradish peroxidase-mimicking DNAzyme (HRP-DNAzyme) where a K(+)-stabilized single stranded DNA (ssDNA) with G-rich sequence was used as the recognition element. HRP-DNAzyme is formed in the presence of K(+) which further allowed a dye oxidization by H_2O_2 into a fluorescent product. The fluorescence intensity showed linear relation to the logarithm of K^+ concentration in the range of 2.5 μM–5 mM. Metal ions, such as Na^+, Li^+, $NH_4{}^+$, Mg^{2+}, and Ca^{2+} did not interfere with detection of K^+ [40].

Graphene oxide (GO) was introduced as an efficient quencher for label-free and sensitive detection of DNA. Probe DNA (pDNA) was mixed with ethidium bromide (EB) and graphene oxide (GO). The interaction between EB and GO led to the fluorescent quenching. Upon target recognition, EB was intercalated into duplex DNA and kept away from GO, which significantly hindered the long range resonance energy transfer (LrRET) from EB to GO and, thus, increased the fluorescence of EB. The changes in fluorescent intensity produced a novel method for sensitivity and specificity detection of the target. Based on the structure switching of aptamers, this strategy could be conveniently extended for detection of other biomolecules, which had been demonstrated by the detection of exonuclease activity [41].

In another approach $NaYF_4$:Yb,Er upconversion nanoparticles (UCNPs) act as donor and gold nanorods (AuNRs) as acceptor under red region in a FRET system for determination of thrombin, Both the NPs and the AuNRs were conjugated with thrombin aptamers which showed the formation of FRET system only when thrombin was added. It was also demonstrated that the linear range was 2.5–90 nM in an aqueous buffer, and 3.75–112.5 nM in spiked human serum samples for thrombin [42]. Similar work was previously reported using graphene FRET system by Li et al. FRET-based sensing system for detecting human platelet-derived growth factor-BB (PDGF-BB) using fluorophore-functionalized aptamers and quencher-carrying strands hybridized in duplex, coupled with streptavidin (SA)-functionalized silver nanoparticles. This AgNP's-FRET sensor showed low fluorescence intensity due to FRET effect; however in the presence of the growth factor, the quencher was displaced leading to increase in the fluorescence intensity signal. A good linear response was obtained when the PDGF-BB concentrations are in the ranges of 100–500 and 6.2–50 ng/mL with the detection limit of 0.8 ng/mL [43].

7.3 Thermal Sensors

Basic Principle of Thermal Sensors

The first law of thermodynamics: Any process in which the internal energy of the system changes is accompanied by absorption or *evolution of heat*. Two properties of heat: Heat is totally nonspecific, and heat cannot be stored. Heat evolved is measured in terms of ΔT or the heat flux. The heat evolved continuously, so thermal sensors are in a *non-equilibrium state* and their signal is obtained from a *steady-state* situation. The heat generated at constant pressure C_p is equal to the enthalpy change ΔH, which is related to the change in temperature ΔT. If C_p is constant over a given temperature range, the enthalpy change can be calculated from such a measurement for any temperature within that range as

$$dT = -\frac{dH}{C_p}.$$

From C_p, we can calculate C_s (thermal capacity).

Push–Pull Principle: Heat flowing in should be balanced with heat flowing out.

$$\frac{d}{dt}(\pm\,\Delta H\text{rxn} + \Delta H\text{el} + \Delta H\text{loss})_T = 0.$$

When no chemical reaction occurs: Reaction heat = 0. When chemical reaction occurs: Reaction heat = Heat supplied. P-P scheme can be used for thermal sensors operating in gas and liquid phases.

These sensors are among the oldest sensors (over 150 years) that generate electrical e.m.f.s (voltages) directly by measuring very small voltages difficult to measure and often amplified before interfacing. However, this can be influenced by noise. These sensors operate on almost the entire range of temperature from near absolute zero to about 2700 °C. No other sensor technology can match even a fraction of this range and can be designed by anyone with minimum skill as these are simple, rugged, and inexpensive made at the sensing site if necessary.

Seebeck discovered that dissimilar metals that are connected at two different locations (junctions) will develop microvoltage if the two junctions are held at different temperatures. This effect is known as the "Seebeck effect"; it is the basis for thermocouple thermometers [44]. Peltier Effect: In 1834, Peltier discovered the inverse of the Seebeck effect, now known as the "Peltier effect": He found that if you take a thermocouple and apply a voltage, this causes a temperature difference between the junctions. This results in a small *heat pump*, later referred to as also known as a *thermoelectric cooler (TEC)* (Fig. 7.40).

Thomson Effect: A current-carrying wire if unevenly heated along its length will either absorb or radiate heat depending on the direction of current in the wire (from hot to cold or from cold to hot).

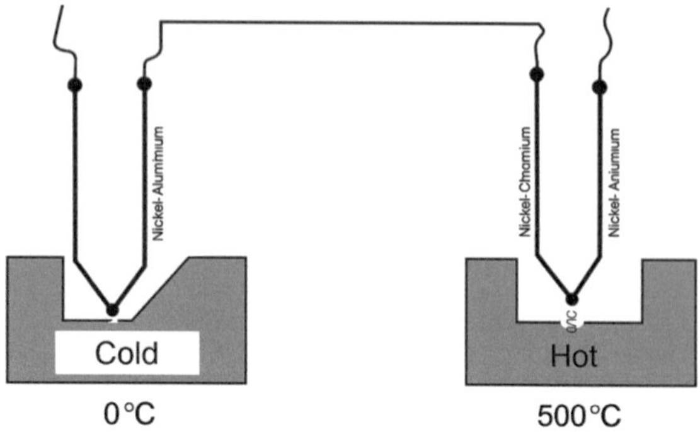

Fig. 7.40 Peltier effect showing a thermocouple applied with a voltage causing temperature difference between the junctions 0 and 500 °C

Thermocouples: A thermocouple is a sensor that measures temperature. It consists of two different types of metals, joined together at one end. When the junction of the two metals is heated or cooled, a voltage is created that can be correlated back to the temperature. Thermocouple alloys are commonly available as wire. A thermocouple probe consists of thermocouple wire housed inside a metallic tube. The wall of the tube is referred to as the sheath of the probe. The tip of the thermocouple probe is available in three different styles. Grounded, ungrounded, and exposed.

Resistive Thermometers: These are sensors used to measure temperature by correlating the resistance of the RTD element with temperature. Most RTD elements consist of a length of fine coiled wire wrapped around a ceramic or glass core often placed inside a sheathed probe to protect it. The RTD element is made from pure material, typically platinum, nickel, or copper. The material has a predictable change in resistance as the temperature changes. They are slowly replacing the use of thermocouples in many industrial applications below 600 °C, due to higher accuracy and repeatability. Resistive Thermometer Devices: RTDs are very linear devices, but not perfect. To calculate the resistance of the RTD, we must pass a current through the device. This creates two problems: Power will be dissipated in the RTD itself creating heat (self-heating). The lead wires will have some resistance that will be included in the resistance attributed to RTD itself. (2) Ω resistance in the leads with an RTD resistance of 107.79 Ω (20 °C) converts to a temperature of 25 °C (5 °C error) $\Delta T = \Delta R/0.392$.

Thermistors: Thermistors are the probes of choice because they are stable (± 0.05 °C/year), inexpensive, sensitive, chemically inert, and small). They are made of high-temperature sintered oxides (such as BaO/SrO) covered with a glass coating. They cover a temperature range from − 80 to + 350 °C and have negative (NTC) or positive (PTC) temperature coefficient [45, 46].

Glucose Thermistor: In this device, enzyme glucose oxidase is immobilized on the thermistor ceramic and surrounded by a thermal shield [47]. The enzyme catalyzes the oxidation of glucose according to the reaction and produces hydrogen peroxide and heat ΔH1. The hydrogen peroxide is eliminated by the following reaction catalyzed by catalase and generating ΔH2 where H1 and H2 are the enthalpy changes during the enzymatic reactions with glucose oxidase and catalase, respectively. The other enzyme tested namely urease catalyzes the hydrolysis of urea according to the following reaction:

$$(NH_2)_2CO + 3H_2O \rightarrow HCO_3^- + 2NH_4^+ + OH^- + H + \Delta H_3.$$

7.4 Mass Sensors

Piezoelectric Effect. The word "piezo" comes from the Greek word "piezen" which translates "to squeeze." The generation of an electrical signal by a dielectric material as it is subjected to a mechanical stress. Rochelle salt was first used in sensors as it has a large piezoelectric effect, however is less stable than other materials. Tourmaline, a silicate mineral that can generate an electric charge when subjected to mechanical pressure or heat. Quartz, a common crystal used in clocks, it is a naturally occurring piezoelectric material. Some biological materials such as bone and DNA also exhibit piezoelectric properties. Man-made piezoelectric materials include Langasite (LGS) a synthetically grown single crystal with superior high-temperature stability and a higher electromechanical coupling factor than quartz, making it valuable for use in extreme environments. Other examples include Lithium Niobate, a man-made single crystal with applications in optical waveguides, sensors, and mobile phones, quartz grown artificially for use in electronics, piezofilm, a thin, flexible film made from the special polymer polyvinylidene fluoride (PVDF) that is artificially polarized, and its flexibility makes it useful for sensors and transducers. Piezoceramics materials include lead, Zirconate, Titanate (PZT), and Bismuth Titanate. Why piezoelectric materials and what are their properties. Piezocrystallites are centro-symmetric cubic (isotropic) before poling and after poling exhibit tetragonal symmetry (anisotropic structure) below the Curie temperature. Above this temperature, they lose the piezoelectric properties.

Quartz Crystal: Crystals are class of materials arranged in a definite, geometric pattern in three dimensions (table salt and sugar are common examples). Quartz crystal is silicon and oxygen arranged in a crystalline structure (SiO_2). SiO_2 is also found abundantly in nature in a non-crystal structure ("amorphous") as sand. Quartz is the material of choice for most piezoelectric sensor applications. Why? It is inexpensive, relatively strong piezoelectric coefficient and its hexagonal crystallographic structure- no center of symmetry.

The unique *piezoelectric* and *converse piezoelectric* properties of *crystalline quartz* allow us to design an electro-mechanical device (mechanical force = electrical signal; electrical field = mechanical deformation). By operating the device at its *mechanical resonance frequency*, we can get a useful electrical signal out of it: The electrical signal will be at the same frequency as the mechanical resonance frequency. If we know how to control the device's *mechanical resonance frequency*, we ought to be able to design a device at any frequency of interest.

Quartz Crystal Microbalance

Resonance: Resonance is the tendency of a system to oscillate with greater amplitude at some frequencies.

Thickness Shear Mode (TSM) Resonators or QCMs

Sauerbrey Equation: The **Sauerbrey equation** was developed by G. Sauerbrey in 1959 as a method for correlating changes in the oscillation frequency of piezoelectric crystal with the mass deposited on it. He simultaneously developed a method for measuring the characteristic frequency and its changes by using the crystal as the frequency determining component of an oscillator circuit. His method continues to be used as the primary tool in QCM experiments for conversion of frequency to mass and is valid in nearly all applications.

Assumptions and Limitations: Because the film is treated as an extension of thickness, Sauerbrey's equation only applies to systems in which the following three conditions are met: The deposited mass must be rigid, the deposited mass must be distributed evenly, the frequency change $\Delta f/f < 0.05$, "Sauerbrey limit" is the upper amount of mass added beyond the equation. 1 Hz shift = 17.7 ng cm^{-2} and dynamic range is 20 μg cm^{-2}. The material added to the QCM will invariably exhibit different mechanical characteristics than quartz. Unified behavior is weak. Mechanical clamping, Damping: is an effect that reduces the amplitude of oscillations in an oscillatory system. The system oscillates (at reduced frequency compared to the *undamped* case) with the amplitude gradually decreasing to zero and Temperature.

Q Factor: The *Q* factor determines the qualitative behavior of simple damped oscillators. A system with low quality factor ($Q < ½$) is said to be overdamped. Such a system does not oscillate at all but when displaced from its equilibrium steady-state output it returns to it by exponential decay, approaching the steady-state value asymptotically. A system with high quality factor ($Q > ½$) is said to be underdamped. Underdamped systems combine oscillation at a specific frequency with decay of the amplitude of the signal. A system with an intermediate quality factor ($Q = ½$) is said to be critically damped. Like an overdamped system, the output does not oscillate, and does not overshoot its steady-state output.

In the context of resonators, *Q* is defined in terms of the ratio of the energy stored in the resonator to the energy supplied by a generator, per cycle, to keep signal amplitude constant, at a frequency (the resonant frequency), f_r where the stored energy is constant with time. *Q* is defined as

$$Q = 2p \times \frac{\text{Energy Stored}}{\text{Energy dissipated per cycle}} = 2pf_r \times \frac{\text{Energy Stored}}{\text{Power Loss}}.$$

For high values of *Q*, the following definition is also mathematically accurate as shown below

$$Q = \frac{f_r}{\Delta f} = \frac{w_r}{\Delta w},$$

where f_r is the resonant frequency, Δf is the half-power bandwidth, $\omega_r = 2\pi f_r$ is the angular resonant frequency, and $\Delta\omega$ is the angular half-power bandwidth. In Sauerbrey equation we assume

$$V \text{ (shear velocity)} = F \text{ (freq)} \times \lambda \text{ (Wavelength)}; \text{ where } \lambda = 2t.$$

Shear velocity of crystal $v = \left(\frac{\mu}{\rho}\right)^{1/2}$ where μ = shear modulus. Shear velocity of metal electrode film $vf = \left(\frac{\mu f}{\rho f}\right)^{1/2}$

$$\rho f v f / \rho v = Zf / Z$$

The ratio of impedance is important in acoustic matching of materials. QCMs with higher ratio result in low ***Q* factor**, which often do not oscillate.

QCM in aerosol (Table 7.2) and suspension phase: Monitoring heterogeneous samples—Aerosol and suspensions. Liquid-aerosol and solid-suspension = sticky particles. Particle diameter plays an important role where increased diameter of particleis equal to increase in contact area which corelates to increase in the inertial force to remove the particle. Therefore, the optimum diameter (D) is inversely proportional to frequency according to

$$D = \frac{\beta}{F0}$$

where β is the particle density.

For QCM, mass sensitivity depends on radial distance. Enhancement of the stickiness: poly siloxane silicone grease. [Concentration of Aerosol] = mass of liquid phase/sample volume. The efficiency of the sensor is calculated using

$$\text{Ec} = \Delta M / M$$

Ec = 1 $\mu g\ m^{-3}$ of air sample can be obtained.

QCM in liquid phase: An additional load of liquid increases the complexity of the sensor performance. Therefore, sensor and interfacial properties must be studied. Increase in complexity of sensor performance. Sensor and interfacial properties must be studied. Liquid is an additional load. Transverse shear waves of the liquid must be taken into account. Energy is dissipated (loss) into the liquid. causing acoustic signal attenuation. If a layer of water moves along the surface, then, velocity of the liquid = δ (damping distance) according to

$$\delta = \left(\frac{KL}{\pi F}\right)^{1/2}$$

Table 7.2 QCM in aerosol phase

Material	Shear mode acoustic wave impedance (10^6 kg s^{-1} m^{-2})	
Aluminum	8.22	0.99
Aluminum oxide	24.6	
Chromium	29	
Copper	20.3	
Gold	**23.2**	2.8
Graphite	2.71	
Indium	10.5	
Nickel	26.7	
Quartz	*8.27*	
Platinum	**36.1**	4.36
Silicon	12.4	
Silver	16.7	

KL = kinematic viscosity.

If liquid layer is stable on the surface the δ = thickness, then

$$\Delta M = \rho Lt.$$

Effects caused in liquid phase include ecessive mass loading change in mechanical properties of selective layer, stress caused by electrode mounting. positioning of sensor in fluid compartment and conductivity. liquid trapping.

Surface Acoustic Wave: A surface acoustic wave (SAW) is an acoustic wave **traveling along the surface** of a material exhibiting elasticity, with amplitude that typically decays exponentially with depth into the substrate. Surface acoustic waves were discovered in 1885 by Lord Rayleigh and are often named after him: Rayleigh waves. A surface acoustic wave is a type of mechanical wave motion which travels along the surface of a solid material. The velocity of acoustic waves is typically 3000 m/s, which is much lower than the velocity of the electromagnetic waves. Surface acoustic wave sensors are a class of microelectromechanical systems (MEMS) which rely on the modulation of surface acoustic waves to sense a physical phenomenon.

The sensor transduces an input electrical signal into a mechanical wave which, unlike an electrical signal, can be easily influenced by physical phenomena. The device then transduces this wave back into an electrical signal. Changes in amplitude, phase, frequency, or time-delay between the input and output electrical signals can be used to measure the presence of the desired phenomenon.

SAW Devices: The operation of the SAW device is based on acoustic wave propagation near the surface of a piezoelectric solid. This implies that the wave can be trapped or otherwise modified while propagating. The displacements decay exponentially away from the surface, so that the most of the wave energy (usually more than 95%) is confined within a depth equal to one wavelength. t (thickness of the surface) > λ of the propagation wave. The surface wave can be excited electrically by means of an interdigital transducer (IDT). A basic SAW device consists of two IDTs on a piezoelectric substrate such as quartz. The input IDT launches, and the output IDT receives the waves. **Principle**: The interdigital transducer consists of a series of interleaved electrodes made of a metal film deposited on a piezoelectric substrate as shown above. The width of the electrodes usually equals the width of the inter-electrode gaps (spacing) giving the maximal conversion of electrical to mechanical signal, and vice versa. The minimal electrode width which is obtained in industry is around 0.3 μm, which determines the highest frequency of around 3 GHz. The commonly used substrate crystals are: quartz, lithium niobate, lithium tantalate, zinc oxide, and bismuth germanium oxide. They have different piezoelectric coupling coefficients and temperature sensitivities. The ST quartz is used for the most temperature stable devices. The wave velocity is a function of the substrate material and is in the range of 1500–4800 m/s, which is 105 times lower than the electromagnetic wave velocity. This enables the construction of a small size delay line of a considerable delay. The input and output transducers may be equal or different. It depends upon the function which the SAW device must perform. Usually, they differ in electrode's overlaps, number, and sometimes positioning.

Mass sensors based on non-piezoelectric materials:

Cantilevers

Bending Mode: In bending mode, the adsorption of material at one surface causes change of surface energy (stress) and bending. One equation is the key to understand the behavior of cantilevers. The first is *Stoney's formula*, which relates cantilever end deflection δ to applied stress σ: where v is Poisson's ratio, E is Young's modulus, L is the beam length, and t is the cantilever thickness. Very sensitive optical and capacitive methods have been developed to measure changes in the static deflection of cantilever beams used in DC-coupled sensors.

Resonant Mode: Another formula relating the cantilever is the spring constant to the cantilever dimensions and material constants. Hooke's law = Force $\propto x$ (distance of spring); $k = Fx$ or δ (deflection) where k is the spring constant according to

$$k = \frac{F}{\delta} = \frac{E\omega t3}{4L3}$$

Here is force and ω is the cantilever width. Resonance frequency can be calculated. Deposition of mass ΔM results in decrease of the resonant frequency (Fig. 7.41).

These modes enable label-free, sensitive detection in physical, chemical, and biological sensing applications.

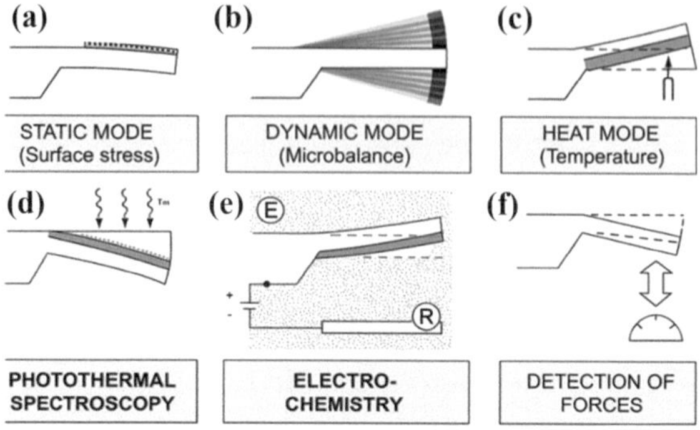

Fig. 7.41 Schematic overview of various operational modes of microcantilever-based sensors: **a** *Static mode*—detects surface stress changes due to molecular binding, leading to cantilever bending. **b** *Dynamic mode*—operates as a microbalance where mass loading alters resonance frequency. **c** *Heat mode*—monitors temperature-induced deflection, applicable in thermal sensing. **d** *Photothermal spectroscopy*—measures laser-induced heating and thermal expansion effects. **e** *Electrochemistry*—uses electrochemical reactions to induce surface stress and deflection in the cantilever. **f** *Detection of forces*—measures intermolecular or mechanical forces via cantilever deflection

Advantages

Operate at frequencies (50 MHz) well below the velocity of acoustic waves in liquid media. Problem of energy dissipation is diminished. No electrical connections. Optical readout (position sensitive detector). Lateral and torsional mode of oscillations. Great flexibility in the design, fabrication, readout and used as arrays for gaseous and biosensing applications [48].

References

1. P. R. Buck and E. Lindner, "IUPAC Recommendation for Nomenclature of Ion-Selective Electrodes," Pure and Applied Chemistry, Vol. 66, No. 12, 1994, pp. 2527–2536.
2. Janata, J. and J. J. Janata, Jiri, and Jií Janata. 'Potentiometric sensors.' Principles of Chemical Sensors (2009): 119–199.
3. Štulík, K. ION-SELECTIVE ELECTRODES | Overview. *Encyclopedia of Analytical Science: Second Edition* 493–498 (2005) https://doi.org/10.1016/B0-12-369397-7/00291-0.
4. De Marco, R. & Clarke, G. ELECTRODES | Ion-Selective Electrodes. *Encyclopedia of Electrochemical Power Sources* 103–109 (2009) https://doi.org/10.1016/B978-044452745-5.00848-0.
5. Somer, G., Kalayci, Ş. & Başak, I. Preparation of a new solid state fluoride ion selective electrode and application. *Talanta* **80**, 1129–1132 (2010).
6. Cheong, Y. H., Ge, L. & Lisak, G. Highly reproducible solid contact ion selective electrodes: Emerging opportunities for potentiometry – A review. *Anal Chim Acta* **1162**, 338304 (2021).
7. Nikolskii, B. P. & Materova, E. A. Solid Contact in Membrane Ion-Selective Electrodes. *Ion-selective electrode reviews* **7**, 3–39 (1985).

8. Grygolowicz-Pawlak, E., Plachecka, K., Brzozka, Z. & Malinowska, E. Further studies on the role of redox-active monolayer as intermediate phase of solid-state sensors. *Sens Actuators B Chem* **123**, 480–487 (2007).
9. Vázquez, M., Bobacka, J., Ivaska, A. & Lewenstam, A. Influence of oxygen and carbon dioxide on the electrochemical stability of poly(3,4-ethylenedioxythiophene) used as ion-to-electron transducer in all-solid-state ion-selective electrodes. *Sens Actuators B Chem* **82**, 7–13 (2002).
10. Hamdani, K. & Cheng, K. L. Polyaniline pH Electrodes. *Microchemical Journal* **61**, 198–217 (1999).
11. Hailin, G. & Yucheng, L. Characterization of a chemoresistor pH sensor based on conducting polypyrrole. *Sens Actuators B Chem* **21**, 57–63 (1994).
12. Korostynska, O., Arshak, K., Gill, E. & Arshak, A. Review on State-of-the-art in Polymer Based pH Sensors. *Sensors* **7**, 3027–3042 (2007).
13. Eddowes, M. J. Response of an enzyme-modified pH-sensitive ion selective device; consideration of the influence of the buffering capacity of the analyte solution. *Sensors and Actuators* **7**, 97–115 (1985).
14. van der Schoot, B. H., Voorthuyzen, H. & Bergveld, P. The pH-static enzyme sensor: Design of the pH control system. *Sens Actuators B Chem* **1**, 546–549 (1990).
15. Makarychev-Mikhailov, S., Shvarev, A. & Bakker, E. New trends in ion-selective electrodes. *Electrochemical Sensors, Biosensors and their Biomedical Applications* 71–114 (2008) https://doi.org/10.1016/B978-012373738-0.50006-4.
16. Dzyadevych, S. V., Soldatkin, A. P., El'skaya, A. V., Martelet, C. & Jaffrezic-Renault, N. Enzyme biosensors based on ion-selective field-effect transistors. *Anal Chim Acta* **568**, 248–258 (2006).
17. Khaled, E., Hassan, H. N. A., Mohamed, G. G., Ragab, F. A. & Seleim, A. E. A. Disposable potentiometric sensors for monitoring cholinesterase activity. *Talanta* **83**, 357–363 (2010).
18. Langer, J. J. *et al.* Polyaniline biosensor for choline determination. *Surf Sci* **573**, 140–145 (2004).
19. Trivedi, U. B. *et al.* Potentiometric biosensor for urea determination in milk. *Sens Actuators B Chem* **140**, 260–266 (2009).
20. Syu, M. J. & Chang, Y. S. Ionic effect investigation of a potentiometric sensor for urea and surface morphology observation of entrapped urease/polypyrrole matrix. *Biosens Bioelectron* **24**, 2671–2677 (2009).
21. Cowell, D. C., Dowman, A. A., Newman, J. D., Pirzad, R. & Rantle, C. The polycrystalline fluoride ion-selective electrode and horseradish peroxidase—an alternative electrochemical biosensor? *Biosens Bioelectron* **7**, 27–38 (1992).
22. Brochot, J. A. & Siddiqi, I. W. Peroxidase-catalysed rupture of the C–F bond as and indicator reaction for enzyme immunoassay: Kinetic determination of human immunoglobulin G, α-fetoprotein and placental lactogen. *Anal Chim Acta* **224**, 329–337 (1989).
23. Özbek, O., Berkel, C., Isildak, Ö. & Isildak, I. Potentiometric urea biosensors. *Clinica Chimica Acta* **524**, 154–163 (2022).
24. Wang, Joseph. *Analytical Electrochemistry*. (Wiley-VCH, 2000).
25. Venton, B. J. & DiScenza, D. J. Voltammetry. *Electrochemistry for Bioanalysis* 27–50 (2020) https://doi.org/10.1016/B978-0-12-821203-5.00004-X.
26. Zhu, Y. C. *et al.* Enzyme-Based Biosensors and Their Applications. *Advances in Enzyme Technology, First Edition* 201–223 (2019) https://doi.org/10.1016/B978-0-444-64114-4.00008-X.
27. Forster, R. J. & Walsh, D. A. VOLTAMMETRY | Organic Compounds. *Encyclopedia of Analytical Science: Second Edition* 217–226 (2005) https://doi.org/10.1016/B0-12-369397-7/00651-8.
28. Bard, A. J. & Faulkner, L. R. *Electrochemical Methods: Fundamentals and Applications*. (John Wiley & Sons, Inc., 2001).
29. Brett, C. M. A. & Brett, A. M. Oliveira. *Electrochemistry: Principles, Methods, and Applications*. (Oxford University Press, 1993).

30. Katz, E. & Willner, I. *Probing Biomolecular Interactions at Conductive and Semiconductive Surfaces by Impedance Spectroscopy: Routes to Impedimetric Immunosensors, DNA-Sensors, and Enzyme Biosensors.*
31. Baldini, F. & Giannetti, A. Optical chemical and biochemical sensors: new trends (Invited Paper) (eds. Byrne, H. J. et al.) 485 (2005). https://doi.org/10.1117/12.610653.
32. Homola, J. Surface Plasmon Resonance Biosensors for Food Safety, in 145–172 (2004). https://doi.org/10.1007/978-3-662-09111-1_7.
33. Giuliani, J. F., Wohltjen, H. & Jarvis, N. L. Reversible optical waveguide sensor for ammonia vapors. *Opt Lett* **8**, 54 (1983).
34. Borisov, S. M. & Wolfbeis, O. S. Optical Biosensors. *Chem Rev* **108**, 423–461 (2008).
35. Arnold, M. A. Enzyme-Based Fiber Optic Sensor. *Anal Chem* **57**, 565–566 (1985).
36. Li, F. & Cui, H. A label-free electrochemiluminescence aptasensor for thrombin based on novel assembly strategy of oligonucleotide and luminol functionalized gold nanoparticles. *Biosens Bioelectron* **39**, 261–267 (2013).
37. Chen, C., Zhao, J., Jiang, J. & Yu, R. A novel exonuclease III-aided amplification assay for lysozyme based on graphene oxide platform. *Talanta* **101**, 357–361 (2012).
38. Hu, P., Zhu, C., Jin, L. & Dong, S. An ultrasensitive fluorescent aptasensor for adenosine detection based on exonuclease III assisted signal amplification. *Biosens Bioelectron* **34**, 83–87 (2012).
39. Huang, Y. *et al.* Label-Free Colorimetric Aptasensor Based on Nicking Enzyme Assisted Signal Amplification and DNAzyme Amplification for Highly Sensitive Detection of Protein. *Anal Chem* **85**, 4423–4430 (2013).
40. Fan, X. *et al.* A novel label-free fluorescent sensor for the detection of potassium ion based on DNAzyme. *Talanta* **89**, 57–62 (2012).
41. Jiang, Y. *et al.* A Graphene Oxide–Based Sensing Platform for The Label-free Assay of DNA Sequence and Exonuclease Activity via Long Range Resonance Energy Transfer. *J Fluoresc* **23**, 697–703 (2013).
42. Chen, H. *et al.* Aptamer-based sensing for thrombin in red region via fluorescence resonant energy transfer between $NaYF_4$:Yb,Er upconversion nanoparticles and gold nanorods. *Biosens Bioelectron* **48**, 19–25 (2013).
43. Li, H. *et al.* Silver Nanoparticle-Enhanced Fluorescence Resonance Energy Transfer Sensor for Human Platelet-Derived Growth Factor-BB Detection. *Anal Chem* **85**, 4492–4499 (2013).
44. Allsop, T. & Neal, R. A review: Evolution and diversity of optical fibre plasmonic sensors. *Sensors (Switzerland)* vol. 19 Preprint at https://doi.org/10.3390/s19224874 (2019).
45. Dinh, T. *et al.* Physical Sensors: Thermal Sensors. *Encyclopedia of Sensors and Biosensors: Volume 1–4, First Edition* **1–4**, 20–33 (2023).
46. Zhang, J. X. J. & Hoshino, K. Mechanical transducers: Cantilevers, acoustic wave sensors, and thermal sensors. *Molecular Sensors and Nanodevices* 311–412 (2019) https://doi.org/10.1016/B978-0-12-814862-4.00006-5.
47. Danielsson, B., Gadd, K., Mattiasson, B. & Mosbach, K. Enzyme thermistor determination of glucose in serum using immobilized glucose oxidase. *Clinica Chimica Acta* **81**, 163–175 (1977).
48. Janata, J. Mass Sensors, in *Principles of Chemical Sensors* 63–98 (Springer US, Boston, MA, 2009). https://doi.org/10.1007/b136378_4.

Chapter 8
Biosensor Technology: Structure, Design, and Fabrication

In recent years, the healthcare industry has seen significant integration of biosensors for diagnostic purposes. These electronic devices measure biological signals and convert them into electrical outputs. A typical biosensor includes a bioreceptor, transducer, analyte, and display unit. The bioreceptor specifically interacts with the analyte, triggering a response that the transducer converts into an electrical signal suitable for digital display and interpretation.

Biosensors have transformed healthcare by enabling non-invasive detection and real-time monitoring of various medical conditions—without requiring blood, serum, or fluid extraction. By detecting specific analytes, they facilitate early diagnosis and empower individuals to monitor their health and respond promptly, reducing reliance on clinical visits. Real-time data access enhances patient autonomy and helps minimize complications.

Given the vast potential of biosensors, ongoing research continues to explore their applications across medical and mobile platforms. Among these, the glucose biosensor is the earliest and most commercially successful example, widely used for diabetes management. Biosensors have also demonstrated diagnostic capabilities for viruses such as influenza, SARS, COVID-19, Hendra, and Nipah, expanding their utility in infectious disease surveillance.

In recent years, biosensors have evolved significantly, catering to different analytes including enzymes, proteins, glucose levels, and diseases such as cancer, diabetes, and cardiac conditions. Of particular concern are cardiac diseases, which have garnered considerable attention due to their high mortality rates worldwide. Biomarkers and enzymes serve as primary analytes for diagnosing such diseases. Biosensors employ various transducer principles, biological receptors, and analytes to accomplish their functions. Biological recognition elements encompass categories such as enzymes, aptamers, antibodies, nucleic acids, cells and tissues, and molecularly imprinted polymers (MIPs).

Enzymes, as biocatalysts, play a crucial role in biosensors. They accelerate reaction rates and offer high stability, specificity, and ease of separation after the reaction.

A. Simonian et al., *Biosensors as Analytical Tools for the 21st Century*,
https://doi.org/10.1007/978-3-032-01726-0_8

Enzymes can be immobilized on receptor surfaces, making them reusable. Aptamers, which adopt a 3D structure upon interacting with target entities, exhibit high affinity and specificity. Their binding characteristics are disrupted even by minor changes in the target molecule. Antibodies, generated by the immune system in response to foreign substances called antigens, display high affinity and specificity toward their target molecules. They are well-characterized and can be produced on a large scale at minimal cost. Proteins, composed of amino acids, enable precise and sensitive detection. Molecular imprinting involves creating template-shaped cavities in polymer matrices with predetermined selectivity and high affinity. These templates can consist of bacteria, bacteriophages, molecules, or biomarkers. Molecularly imprinted polymers (MIPs), which mimic natural recognition entities like antibodies and biological receptors, offer advantages such as stability, cost-effectiveness, and ease of preparation.

Based on the employed transducer, biosensors can be further classified into electrochemical, optical, and mass-based biosensors. These biosensors operate on different detection principles. Optical-based biosensors utilize phenomena like total internal reflection, the evanescent wave, and surface Plasmon resonance to facilitate detection. For instance, total internal reflection occurs when a light beam propagates from a medium with a higher refractive index to one with a lower refractive index, resulting in internal reflection and the generation of an evanescent wave. Surface Plasmons are electromagnetic excitations that arise from the interaction of light photons with free electrons at the interface between a dielectric and a metal. In surface Plasmon resonance (SPR) biosensors with angular modulation, monochromatic light at varying angles of incidence is used to excite surface Plasmons. Changes in the refractive index of the sensing layer lead to shifts in the resonance angle.

Electrochemical sensors detect chemical reactions by monitoring changes in the electrical properties of a solution. These sensors typically employ a three-electrode system consisting of a working electrode, a counter electrode, and a reference electrode. The working electrode serves as the transducer where the reaction occurs, while the counter electrode completes the circuit by applying current to the working electrode. The reference electrode is typically set at a stable potential. Subcategories of electrochemical sensors include conductometric biosensors, potentiometric biosensors, and amperometric biosensors. Amperometric sensors measure changes in current resulting from the reduction or oxidation of electroactive species on the electrode surface. Conductometric biosensors rely on changes in electrical conductivity of the electrolyte solution due to the production or consumption of ionic species during metabolic processes.

Mass-based biosensors exhibit changes in oscillating frequency with increasing sensor mass. They are categorized into piezoelectric quartz, surface acoustic, magnetoelastic, and microcantilever biosensors. The underlying principle is that when a heavy metal ion binds to the receptor, the mass changes, leading to a reduced frequency of oscillation.

Biosensors find applications in various fields such as drug delivery, security, food safety, health monitoring, agriculture, and medical diagnosis. Notably, they have facilitated rapid and convenient point-of-care detection in the biomedical field, reducing the need for hospitals and medical practitioners in the diagnostic process. The diverse applications of biosensors emphasize the significance of investigating current research trends and drawing upon knowledge from previous studies by renowned authors.

8.1 Classical Design (Electrochemical and Optical)

The development of electrochemical biosensors has significantly contributed to various fields, including healthcare, environmental monitoring, and food safety. Glucose biosensors, among the first electrochemical biosensors discovered, are widely used for monitoring blood glucose levels in diabetic patients. However, these biosensors often face challenges such as unstable enzyme activity and inhomogeneity, necessitating frequent calibration. To address these limitations, researchers have explored alternative biomolecules with differential electrochemical properties, leading to the discovery of new types of glucose biosensors.

Non-enzymatic biosensors, using synthetic materials instead of proteins, have gained attention due to their stability and selectivity. Different biomolecules with unique electrode stability and selectivity have paved the way for the development of electrochemical biosensors for various purposes. Ferroceneboronic acid (FcBA) and ferrocene (Fc)-modified boronic acids have been reviewed as promising candidates for biosensor development. FcBA and its derivatives can bind to sugar residues, forming cyclic boronate ester bonds and exhibiting different redox properties compared to free FcBA. These properties make them suitable for electrochemical detection, including the measurement of hydrocarbon chains in the polypeptide chain of HbA1c for diabetes diagnosis. Immobilizing FcBA derivatives on electrode surfaces requires further research to improve the performance of FcBA-based electrochemical sensors.

Another important application of electrochemical biosensors is the assessment of antioxidants and reactive oxygen species in physiological systems. Uric acid, a primary end product of purine metabolism, is often detected to diagnose various clinical abnormalities or diseases. Developing a cost-effective and sensitive method for uric acid measurement is crucial. Electrochemical-based approaches similar to glucose biosensors have been considered, but the resemblance of uric acid oxidation to ascorbic acid poses challenges in terms of sensitivity. Scientists have developed amperometric detection-based biosensors that can measure both reduction and oxidation potentials, overcoming this challenge. Immobilization or screen printing of the enzyme on electrodes or nanomaterial-based electrodes has been explored for the development of disposable, selective, cost-effective, and sensitive uric acid biosensors.

Hormone measurements using electrochemical biosensors have also shown promise, but further detailed reviews are needed in this area. Targeting nucleic acids, particularly miRNA expression as a biomarker for disease diagnosis and gene therapy, has shown potential. Electrochemical biosensors based on guanine oxidation have been developed for label-free detection of miRNA, showing promise in this field.

Environmental monitoring is another important application of biosensor technology. Rapid identification of pesticide residues is crucial to prevent health hazards, but traditional methods have limitations in terms of complexity and time-consuming procedures. Enzyme-based biosensors, particularly acetylcholinesterase (AChE) inhibition-based biosensors, have been developed for analyzing pesticides in the environment, food safety, and quality control. Recent advancements in AChE inhibition-based biosensors, including immobilization methods, have improved rapid analysis. Piezoelectric biosensors have also been developed for detecting the impact of organophosphate and carbamate pesticides in the environment. Additionally, electrochemical biosensors using nanomaterials or quartz/silica have revolutionized the detection of organochlorine pesticides and their effects on aquatic ecosystems.

Overall, progress in electrochemical biosensor technology has paved the way for advancements in healthcare, environmental monitoring, and food safety. The selection of receptors, transduction techniques, and screening strategies are important factors to consider for the development and application of biosensors in these fields.

In the quest for simple, swift, and ultra-sensitive biosensors, various materials have been explored, including gold, carbon-based materials, silica, quartz, and glass. These materials play a crucial role in enhancing the performance of biosensors in environmental and biomedical applications. For instance, the incorporation of gold nanoparticles or quantum dots through microfabrication techniques has led to the development of highly sensitive and portable cytochrome P450 enzyme biosensors for specific purposes.

Fiber optic chemical sensors have also gained significant relevance in drug discovery, biosensing, and biomedicine. Recently, hydrogels have emerged as promising materials for immobilization in fiber optic chemistry, particularly in DNA-based sensors. Hydrogels, such as polyacrylamide, are hydrophilic cross-linked polymers that can be shaped into various forms for immobilization, ranging from thin films to nanoparticles. The 3D structure of hydrogels allows for a high loading capacity of sensing molecules, offering advantages such as entrapment, controlled release, analyte enhancement, and DNA protection. The optical transparency of hydrogels facilitates visual detection, making them a convenient substrate for DNA immobilization compared to other materials. Methods for immobilizing DNA biosensors in monolithic polyacrylamide gels and gel microparticles have been developed, representing technical advancements in the field of biosensor technology.

In addition to hydrogels, electrochemical oxidation of hydrazine has enabled single molecule detection in DNA biosensors. This technique has paved the way for highly sensitive and specific DNA detection, further enhancing the capabilities of electrochemical biosensors.

Overall, the utilization of various immobilizers, such as gold, carbon-based materials, silica, quartz, glass, and hydrogels, in combination with advanced fabrication techniques, has significantly contributed to the development of simple, swift, and ultra-sensitive biosensors for diverse applications in environmental monitoring and biomedicine.

8.2 MEMS, Lab-on-Chip, and Microfluidics

Understanding these micro/nanoscale properties and their impact on fluid flow is crucial for designing and optimizing microfluidic devices. Advances in microfabrication technologies have enabled the development of various microfluidic functionalities, such as micropumps, micromixers, particle concentrators, and injection systems (such as nanoneedles). These advancements have opened up new possibilities for applications in chemical analysis, synthesis, and biomedical research, where precise manipulation of ultra-small volumes of liquids is required.

The formula mentioned below relates the applied pressure (P) with the conduit radius (r) for a constant flow rate (Q) in microfluidic devices: $Q \propto r^4$.

This relationship highlights the dominant surface force effects and friction in microfluidic systems due to the increased surface-to-volume ratio compared to macro-scale devices. As the conduit radius decreases in microchannels, the required pressure for pressure-driven flow increases to maintain a constant flow rate. This is because the surface forces become more pronounced at the micro/nanoscale, leading to higher resistance and friction within the fluid flow.

It is important to note that this formula represents a simplified relationship and assumes certain conditions, such as laminar flow and constant fluid viscosity. In practice, other factors such as channel geometry, surface properties, and fluid properties can also influence the flow behavior in microfluidic devices.

Specifically, if the conduit radius is reduced to half, the pressure required to sustain the same flow rate would need to be increased by a factor of 16.

To address the challenges associated with high voltage operation, electrolysis, bubble generation, and pH gradient in Direct Current Electroosmosis (DCEO), Alternating Current Electroosmosis (ACEO) has been explored as a new type of electroosmotic phenomenon for fluidic manipulation at the microscale.

ACEO involves the application of an alternating electric field to induce fluid motion. Unlike DCEO, ACEO operates at lower voltages and does not suffer from electrolysis-related issues or bubble generation. It provides an alternative method for achieving fluidic manipulation and control in miniaturized biochemical devices.

In the context of microcantilever particle trapping, ACEO can be integrated to enhance particle manipulation capabilities. Microcantilevers, which are miniature flexible beams, can be used to trap and manipulate particles in microfluidic systems. By incorporating ACEO, the fluid flow around the microcantilever can be controlled, allowing for precise particle trapping and positioning.

Overall, ACEO offers a promising approach for fluidic manipulation in microfluidic devices, providing advantages over traditional DCEO methods. Its integration with microcantilever particle trapping demonstrates the potential for enhanced functionalities and applications in the field of miniaturized biochemical devices.

Microfluidics is a multidisciplinary field that focuses on shrinking fluid handling systems to a microscale. This concept of miniaturized lab-on-a-chip devices or micro-total analysis systems (μ-TAS) has gained significant interest across various industries. In microfluidics, transport, mixing, reactions, separations, and manipulation

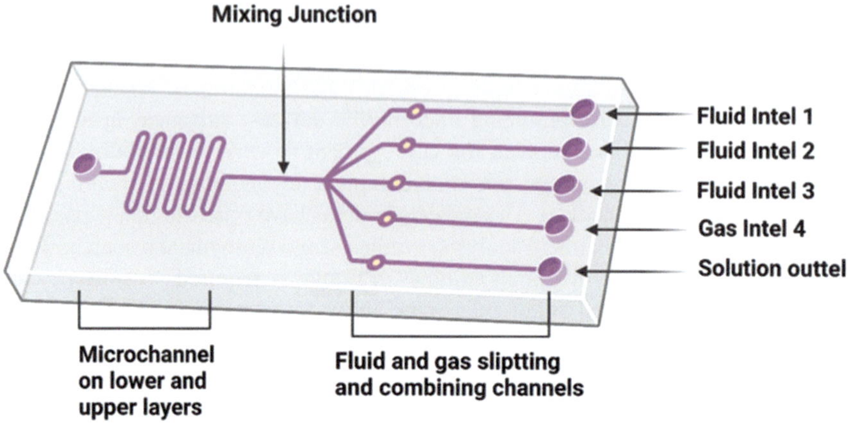

Fig. 8.1 Schematic diagram of a microfluidic chip illustrating the design of fluid and gas input channels and their mixing. The system includes four inlets (fluid inlet 1–3 and gas inlet 4) that lead into a central mixing junction. the device integrates microchannels on both lower and upper layers to enable precise control of fluid and gas flow, including splitting and recombination. The combined output exits through the solution outlet, supporting applications in controlled mixing and microreaction environments [1]

of chemicals and particles are performed on a much smaller scale compared to traditional engineering technologies, as shown in Fig. 8.1.

A functional microfluidic chip should be capable of performing various tasks such as transportation, mixing of reactants, and sample treatment such as concentration and sorting. Unlike the microelectronics industry, which primarily focuses on transistor size reduction, microfluidics aims to explore new fluid phenomena at the micro/nanoscale, enabling more advanced fluid handling capabilities.

One promising type of micropump used in microfluidics is driven by electroosmosis (EO). EO pumps rely on electric fields and do not have any moving parts. They leverage the dominant surface force in microchannels, which is a suitable choice for pumping liquids due to the high surface-to-volume ratio.

The field of microfluidics has turned out to be more complex than initially anticipated. The most efficient mechanisms for fluid or sample manipulation can vary depending on factors such as sample conductivity, pH value, and sample size. It is also essential to consider any potential side effects associated with these manipulation techniques. Therefore, the development of microfluidics has not followed a similar trajectory as the microelectronics industry.

Lab-on-a-chip (LOC) devices have achieved commercial success in biological applications such as electrophoretic separations and DNA sequencing, where DC electrokinetics or DC electric fields are used to manipulate fluids and particles. However, extending DC electrokinetics to more fluidic functions faces challenges due to the high voltage requirements. On the other hand, AC electrokinetics shows great potential in the field of life sciences. With its ability to manipulate particle and fluid motion at the microscale using low voltages, it aligns well with the requirements of lab-on-a-chip systems.

In this chapter, the combination of sample transportation and particle detection in a microfluidic chip is discussed. Biased ACEO micropumping has been experimentally validated for sample transportation, offering a micropump operated with lower AC voltage compatible with lab-on-a-chip systems. For particle detection, effective techniques for detecting micro/nanoscale particles are necessary. Additionally, the integration of a microcantilever with the microfluidic device has been envisioned to enable particle trapping and validation techniques. This integration has the potential to significantly advance AC electrokinetics and its applications in microfluidics.

8.3 Multiplexed Detection

Multiplexing in the context of biosensors refers to the simultaneous detection of multiple biomarkers in a single diagnostic test. This approach is valuable in various medical applications. While single-biomarker detection has its uses, it may not provide sufficient information for clinical diagnosis or disease tracking in many cases. Monitoring the concentrations of multiple biomarkers can be essential for identifying the presence or severity of certain diseases.

Multiplexed sensors offer the capability to detect a panel of discriminative biomarkers simultaneously, leading to improved detection accuracy and enabling early disease diagnostics. Therefore, the development of next-generation sensors has focused on incorporating multidetection approaches.

The levels of discriminative biomarkers often vary at different stages of a particular disease. Cardiovascular diseases serve as a common example where the response to different drugs can be influenced by genetic variations (polymorphisms) among patients. Similarly, certain biomarkers may be shared between different cancer types and benign tumors. Precise identification of specific cancer types often requires the detection of multiple biomarkers. For instance, breast cancer patients typically exhibit significantly higher levels of several biomarkers compared to patients with benign breast tumors.

Despite significant advancements in medical treatments, there is still a high demand for advanced and practical medical services. Delayed diagnosis and treatment, particularly in areas with limited medical resources, can lead to worsening conditions. Additionally, it is crucial to extract as much information as possible from limited sample volumes during diagnosis. In this context, multiplexed bio-(chemical) sensors offer an evident and attractive approach, especially when dealing with limited sample volumes.

By enabling the simultaneous detection of multiple biomarkers, multiplexed biosensors have the potential to enhance diagnostic accuracy, enable early disease detection, and optimize medical services, even in resource-limited settings.

Multiplex detection of protein or peptide biomarkers in human samples for clinical diagnosis poses a challenge due to the presence of diverse proteins and peptides in the sample matrix, leading to potential cross-reactivity. However, advancements in technology have facilitated the adaptation of certain immunoassays for multiplex detection in point-of-care settings.

Lateral Flow Assays (LFAs) are widely used for the detection of protein and peptide biomarkers. They employ detection techniques such as fluorescent immunoassays, chemiluminescence immunoassays, and colorimetric immune assays. LFAs offer rapid results and relatively low fabrication costs compared to other diagnostics. While they have lower sensitivity compared to molecular diagnostic tests, LFAs meet the WHO ASSURED criteria and can be equipment-free or accompanied by a simple reader with a digital interface. The subjective interpretation of colorimetric readouts can be addressed by utilizing a reader, promoting objective and accurate result interpretation, thus meeting the REASSURED criteria.

Enzyme-linked immunosorbent assays (ELISAs) are highly sensitive methods for biomarker detection but are prone to interferences. Developing a multiplex ELISA test is challenging due to these interferences. Spatial multiplexing approaches, such as wells and microarrays, are used to overcome this challenge. ELISAs require multiple wash steps to avoid false positives, making it difficult to automate them for point-of-care diagnostics in limited resource settings. Additionally, lengthy incubation periods are necessary to avoid false negatives, further complicating the adaptation of ELISAs to meet the REASSURED criteria.

Several multiplex immunochromatographic LFAs have been developed for rapid differential diagnosis, such as the BinaxNOW influenza A and B card 2 by Abbott, Becton and Dickinson's (BD) Veritor™ Flu A + B with analyzer, Acucy influenza A and B test by Sekisui Diagnostics, and Quidel's Sofia 2 Flu + SARS antigen FIA test. These tests are accompanied by portable, battery powered readers or analyzers, making them suitable for limited resource settings and meeting the REASSURED criteria.

In summary, the trend in LFA diagnostics is to accompany them with readers and real-time connectivity, enabling rapid adaptation and meeting the criteria for point-of-care diagnostics in various settings.

8.4 Nanostructures in Biosensing

Advancements in nanotechnology have played a significant role in the progress of biosensor development, as well as various other scientific domains. Nanoscale materials possess distinct properties compared to their bulk counterparts, making them highly advantageous in biosensor applications and causing a revolution in this field, as shown in Fig. 8.2. These nanomaterials meet the requirements of transducer materials by exhibiting selectivity, detection capabilities, non-toxicity, biocompatibility, reversibility, fast response, and sensitivity. One of the unique aspects of nanomaterials is their ability to be synthesized in different sizes, shapes, distributions, and compositions, making them ideal for biosensor applications. The integration of nanotechnology into biosensor design is crucial as it ensures superior optoelectronic, electrical

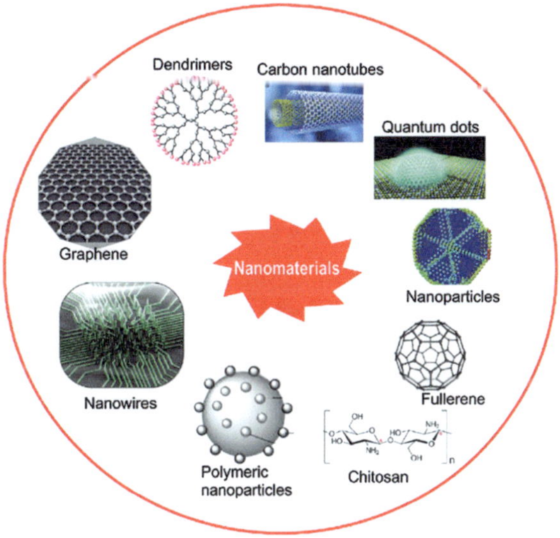

Fig. 8.2 Overview of various types of nanomaterials used in biomedical and technological applications. The figure includes representations of dendrimers, carbon nanotubes, quantum dots, nanoparticles, fullerenes, chitosan, polymeric nanoparticles, nanowires, and graphene, illustrating the structural diversity and functional versatility of nanomaterials [5]

conductivity, catalytic activity, and biocompatibility properties, thanks to their high surface-to-volume ratio.

Due to their exceptional multifunctional properties, researchers have increasingly focused on utilizing diverse nanomaterials in biosensors. Numerous techniques have been reported for fabricating electrodes to develop highly sensitive, selective, and rapid nanosensors in biosensing applications. These methods include employing nanocomposite structures, indium tin oxide (ITO)-polymer, conductive polymer nanoparticles, screen-printing water-based carbon ink, surface molecular imprinting, inkjet printing, recrystallization, injection molding, and charge transfer. Nanomaterials, with their large specific surface area and homogeneous particle distribution, offer excellent detection sensitivity. Notably, graphene, graphene oxide (GO), and reduced graphene oxide (rGO) metal oxide nanocomposites have emerged as favored nanomaterials for producing electrochemical sensors. In recent years, extensive research has been conducted on graphene and graphene-based nanocomposites to enhance biosensor performance. For instance, Eftekhari-Sis et al. [2] developed a novel nanobiosensor using GO/5-carboxy fluorescein-labeled DNA to monitor mutation in exon 19 of the EGFR gene in lung cancer. Bao et al. [3] proposed an effective electrochemical biosensor based on 3D graphene/copper oxide nanoflowers for the detection of malathion, achieving a low detection limit of 0.31 ppt within a concentration range of 1 ppt–15.555 ppb, attributed to the excellent conductivity and adsorption properties of the CuO NFs nanocomposite electrode. Additionally, Taşaltın [4] designed a non-enzymatic PAN: β-rhombohedral borophene-based electrochemical biosensor for glucose detection. This biosensor, combining PAN: β-rhombohedral borophene as a two-dimensional (2D) material, exhibited a low limit of detection (0.099 mM) within a concentration range of 1.5–12 mM and a rapid response time (30 s) due to the electrochemical oxidation of glucose.

To date, nanobiosensors based on 0D–2D nanomaterials have demonstrated high sensing performance in various forms, such as polymer-based, metal-based, carbon-based, and composite-based nanosystems.

8.5 Artificial Intelligence in Biosensing

Machine learning (ML), as part of the artificial intelligence, has proven to be highly effective in processing large sensing data sets, particularly in the context of complex matrices or samples. ML offers several benefits in biosensor applications, including the ability to derive meaningful analytical results from noisy and low-resolution sensing data that may overlap significantly. Furthermore, ML methods can uncover hidden relationships between sample parameters and sensing signals through data visualization and explore interrelations between signals and bioevents. ML can analyze raw sensing data from biosensors in several ways:

1. Categorization: ML algorithms can sort sensing signals into various categories based on the target analyte.
2. Anomaly detection: Biosensors are often influenced by sample matrices and operating conditions, leading to potential contamination. ML can assess the signal and determine if it appears appropriate. It can also address variations in sensor performance caused by biofouling and interferences in real samples.
3. Noise reduction: Sensing signals inherently include noise. ML models can be trained to distinguish between the signal and the noise, considering changes in biosensor signals over seconds or minutes and subsecond interference such as electrical noise.
4. Object identification and pattern recognition: ML algorithms can discover latent objects and patterns within sensing data, enabling easy and effective interpretation.

ML can directly assist biosensor readout, providing automated, accurate, and rapid results, which is crucial for on-site detection or diagnosis. For example, the Orringer group developed a CNN algorithm-assisted optical imaging method that predicts diagnostic results, offering automated readout within 150 s. However, the interpretation of images by pathology experts may still require approximately 30 min.

Furthermore, ML has been utilized to design more desirable biosensors. This section provides an overview of the general types and implementation procedures of advanced ML algorithms. ML is defined as a system or computer program capable of acquiring knowledge by extracting features from raw data. This knowledge can then be utilized to make informed decisions in addressing real-world problems. In the biosensor field, ML serves as a tool for data processing and analysis, encompassing tasks such as feature extraction and prediction of analyte species and concentrations. ML can be categorized into supervised learning and unsupervised learning.

Supervised learning involves training ML algorithms using input data along with their corresponding target outputs. During the training phase, the algorithms make

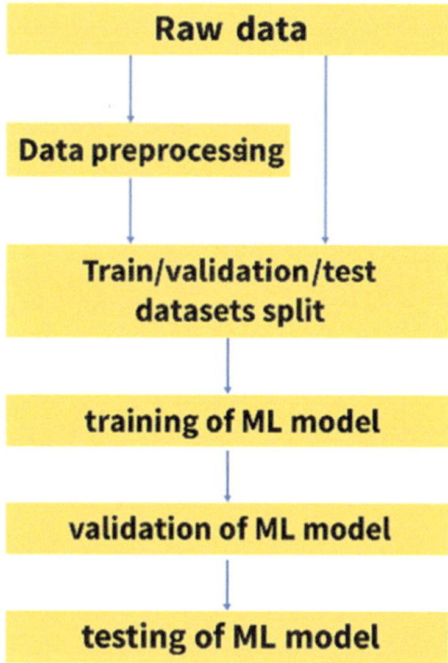

Fig. 8.3 Workflow diagram for machine learning model development. The process begins with raw data collection, followed by data preprocessing. The data are then split into training, validation, and test sets. The machine learning (ML) model is trained on the training set, validated using the validation set, and finally tested on the test set to evaluate its performance

predictions on the input data set and use the true values provided to improve predictive accuracy until it reaches an acceptable level. Supervised learning is commonly employed for classification and regression tasks and has shown significant progress, especially in spectrometric biosensors. Unsupervised learning, on the other hand, deals with situations where labeled training data with target outputs is unavailable. The goal is to determine the data set's distribution in the input space (density estimation) or identify a set of similar examples within the input data set (clustering). The k-Means clustering algorithm is a well-known example of unsupervised learning.

Both sequential data sets (e.g., acquired by electrical and spectral biosensors) and image data sets (e.g., acquired by colorimetric and fluorometric biosensors) exist within the biosensor field. Designing an appropriate ML model based on the data set and analysis purpose (qualitative identification, quantitative detection) is of paramount importance. The workflow described in Fig. 8.3 should be implemented once the ML architecture is specifically designed for a given biosensor.

Preprocessing of raw sensing data is typically required. General preprocessing methods include derivatives, denoising, Fourier transform, and others. System-specific preprocessing methods may involve data compression, elimination of baseline drifts, normalization, transformations, and so on. The application of preprocessing methods can significantly impact the overall performance of ML models.

References

1. Pattanayak, P. *et al.* Microfluidic chips: recent advances, critical strategies in design, applications and future perspectives. *Microfluid. Nanofluidics* **25**, 1–28 (2021).
2. Eftekhari-Sis, M. B., Aliabad, A. & Karimi, F. Graphene oxide based nano-biosensor for the detection of deletion mutation in exon 19 of EGFR gene, leading to lung cancer. *Mater. Lett.* **183**, 441–443 (2016).
3. Bao, J. *et al.* 3D graphene/copper oxide nano-flowers based acetylcholinesterase biosensor for sensitive detection of organophosphate pesticides. *Sensors Actuators B Chem.* **279**, 95–101 (2019).
4. Taşaltın, C. Glucose sensing performance of PAN: β-rhombohedral borophene based non-enzymatic electrochemical biosensor. *Inorg. Chem. Commun.* **133**, (2021).
5. Nagraik, R. *et al.* Amalgamation of biosensors and nanotechnology in disease diagnosis: Mini-review. *Sensors Int.* **2**, 100089 (2021).

Chapter 9
Milestones in Biosensor Applications

In the previous sections, we have discussed the attributes of biosensors, including their types and basic principles. Now, let's focus on their broad implications, which are highly valued not only by industrialists, academicians, and the general public but also by environmentalists, physicians, and medical consultants. The smartness of biosensors arises from their high specificity and sensitivity, reusability, affordability, ease of use, and reliance on fundamental parameters.

Biosensors have found applications across diverse domains, including environmental studies, biotechnology, the food industry, healthcare, and the marine domain. In these fields, biosensors offer numerous advantages over traditional methods, resulting in improved efficiency and effectiveness. They provide more accurate and efficient environmental monitoring, allowing for better decision-making and resource management. Biosensors have revolutionized biotechnology and the food industry by enabling rapid and precise detection of contaminants, ensuring the safety and quality of products.

In healthcare, biosensors have become invaluable tools for diagnostics, disease monitoring, and personalized medicine. Their high sensitivity and specificity enable early disease detection, leading to timely interventions and improved patient outcomes. Biosensors also offer real-time monitoring capabilities, allowing for continuous tracking of vital signs and biomarkers, which is particularly beneficial in critical care settings.

In the marine domain, biosensors play a crucial role in monitoring and preserving the health of marine ecosystems. They provide accurate and real-time data on water quality, pollutant levels, and the presence of harmful substances, helping to mitigate environmental damage and protect marine life.

Overall, biosensors offer significant advantages over traditional methods across various domains. Their wide-ranging impact is highly valued by diverse stakeholders, as they contribute to advancements in industry, academia, public health, environmental conservation, and beyond.

A. Simonian et al., *Biosensors as Analytical Tools for the 21st Century*,
https://doi.org/10.1007/978-3-032-01726-0_9

The use of biosensors for the detection and quantification of heavy metal ions has emerged as a critical application. Heavy metal pollution poses a serious threat to human health, as these ions are persistent in the environment and tend to bioaccumulate. Traditional methods—such as cold vapor atomic absorption spectrometry and inductively coupled plasma mass spectrometry—have been widely used but are limited by the need for highly trained personnel, costly infrastructure, and specialized equipment. Additionally, these techniques are generally restricted to laboratory settings. Biosensors have overcome the limitations of traditional methods by offering high specificity and sensitivity, affordability, ease of use, portability, and real-time signal processing. They provide an efficient and accessible means of detecting and quantifying heavy metal ions, effectively replacing more complex analytical techniques.

Smart biosensors compatible with both protein-based and whole-cell-based systems have proven valuable for heavy metal ion analysis. By leveraging the selectivity of biological components—such as enzymes or antibodies—these sensors specifically bind to target metal ions and produce measurable signals. The rapid, real-time processing of these signals enables immediate detection and quantification. Their cost-effectiveness and user-friendly design make biosensors ideal for diverse applications, including on-site monitoring and field-based analysis. Furthermore, the portability of biosensor devices allows for their deployment in diverse settings, including industrial sites, environmental monitoring stations, and remote locations where conventional methods may be impractical. By providing a reliable and efficient alternative to laboratory-bound techniques, biosensors have revolutionized the detection of heavy metal ions.

Overall, the application of biosensors for the detection and quantification of heavy metal ions represents a critical advancement that addresses the shortcomings of traditional methods. Their high specificity, sensitivity, affordability, ease of use, portability, and real-time signal processing make biosensors an ideal solution for monitoring heavy metal pollution—contributing to the protection of both human health and the environment.

According to projections, approximately 48.3 million people in the USA are expected to be affected by diabetes mellitus by 2050. As the most common endocrine disorder impacting carbohydrate metabolism, diabetes poses serious health challenges for developed societies and can lead to life-threatening complications. The rising prevalence of this condition is largely attributed to increasingly sedentary lifestyles.

To manage their condition and monitor glucose levels, individuals with diabetes commonly use glucometers (blood glucose monitors). These commercially developed devices help users regulate dietary habits and address health concerns. Self-regulation and self-monitoring of blood glucose (SRBG/SMBG) have become fundamental strategies in effective diabetes management. The primary goal of SRBG/SMBG is to help patients achieve and maintain normal blood glucose concentrations, thereby reducing the risk of associated complications such as retinopathy, nephropathy, neuropathy, stroke, and coronary artery disease. By regularly monitoring their blood glucose levels, individuals with diabetes can make informed decisions about their diet, exercise, and medication to effectively manage their condition and minimize the risks associated with diabetes.

The introduction of glucometers and the widespread adoption of self-regulation and blood glucose monitoring have significantly improved diabetes management. These innovations empower individuals to take control of their health, enabling timely interventions and helping to prevent or reduce the complications associated with the disease.

Ongoing efforts aim to enhance the accuracy, convenience, and user-friendliness of glucose monitoring devices, making self-management more seamless and effective. Furthermore, advances in biosensor technology offer promising prospects for the development of non-invasive glucose monitoring methods, minimizing the need for frequent finger pricks and enhancing the quality of life for individuals living with diabetes. Self-regulation and monitoring of blood glucose levels play a critical role in managing diabetes and preventing related complications. Glucometers have transformed diabetes care by empowering individuals to track their glucose levels and make informed decisions about their lifestyle and treatment. Ongoing advancements in this field are expected to further enhance disease management and improve health outcomes for individuals living with diabetes.

In recent years, the development of biosensors using nanotechnology has significantly impacted cancer detection, diagnosis, treatment, and disease management [1].

Traditionally, cancer is often detected at later stages when it has already metastasized, making it challenging to diagnose and treat effectively. However, nanotechnology-based biosensors have effectively addressed this limitation by improving the detection of metastatic cancers and ultimately enhancing patient survival rates. Nanomaterials, including liposomes, dendrimers, buckyballs, and nanotubes, have been utilized as imaging agents, enabling more sensitive and accurate quantification of cancerous tissues. These small yet powerful biosensors offer affordability, ease of use, patient-friendliness, and the ability to access and identify cancer markers, resulting in early cancer detection.

The use of nanotechnology in cancer biosensors has garnered widespread appreciation from both medical consultants and patients alike. These advanced biosensors provide a valuable tool for healthcare professionals in diagnosing and monitoring cancer, while also offering patients a more efficient and effective approach to cancer detection. By leveraging the unique properties of nanomaterials, biosensors have revolutionized cancer diagnostics and have the potential to significantly improve patient outcomes.

It is important to highlight that while nanotechnology-based biosensors hold great promise, ongoing research and development in this field are necessary to further refine and enhance their capabilities. Continued advancements in nanotechnology and biosensor design will contribute to even more sensitive, specific, and reliable cancer detection methods, ultimately leading to improved treatment outcomes and better overall management of this devastating disease.

In the current era, personalized medicine has emerged as a prominent trend, emphasizing the importance of tailoring medical treatments to individual patients. The key principle of personalized medicine is to administer the right drug, at the right dosage, to the right person, and at the right time. This approach recognizes that each person has a unique genetic makeup, lifestyle factors, and health conditions, which can influence the effectiveness and safety of medical treatments.

One critical aspect of personalized medicine is medication adherence, ensuring that patients take their prescribed medications correctly and consistently. To address this challenge, researchers have developed medicine adherence biosensors, which take the form of capsules or similar devices. These biosensors are designed to monitor and track medication intake in a non-invasive and convenient manner.

Medicine adherence biosensors offer several benefits to patients. They provide real-time feedback on medication usage, helping individuals stay on track with their treatment plans. By monitoring medication adherence, these biosensors can alert patients and healthcare providers if doses are missed or if adjustments are needed. This technology promotes patient empowerment and engagement in their own healthcare, enabling them to make informed decisions about their treatment.

Furthermore, medicine adherence biosensors contribute to improved healthcare outcomes. They enhance medication effectiveness by ensuring that patients receive the prescribed dosage at the right time. By promoting adherence, these biosensors can help prevent treatment failures, complications, and hospitalizations. Additionally, they facilitate data collection, allowing healthcare providers to gather valuable insights into patient behavior and treatment response, which can inform personalized treatment plans.

Although medicine adherence biosensors hold great promise, their widespread adoption and integration into healthcare systems are still ongoing processes. As technology continues to advance, these biosensors are expected to become more sophisticated, reliable, and user-friendly. Additionally, addressing concerns related to privacy, data security, and affordability will be crucial to ensure their successful implementation.

Medicine adherence biosensors play a pivotal role in personalized medicine by promoting patient adherence to prescribed medications. These innovative devices contribute to better treatment outcomes, patient empowerment, and improved healthcare management. As personalized medicine continues to evolve, medicine adherence biosensors are poised to become an integral part of patient care, facilitating the delivery of precise and effective treatments to individuals.

Biosensor technology has indeed advanced significantly, and its application in medical diagnostics has driven continuous evolution. The advantages of biosensors in medical diagnostics include ease of use, high sensitivity, capability for multiple analyses, and integration with other technologies such as artificial intelligence, machine learning, big data analytics, and the Internet of Things (IoT). These integrations enable real-time monitoring, diagnosis, prediction, and decision-making, enhancing the potential of biosensors in healthcare.

Implantable biosensors have emerged as the next-generation biosensors for personalized healthcare, allowing continuous monitoring and precise diagnostics.

However, there is still room for improvement to make biosensing devices even more precise and effective as diagnostic tools. Areas for improvement include simplicity, affordability, sensitivity, specificity, accuracy, multiplex analysis of multiple biomarkers, reduction in reagent consumption, and reusability of biological sensing elements. Additionally, making biosensors configurable for continuous monitoring would be beneficial.

While biosensors have found applications in glucometers, ECG management, cancer prediction, biomarker identification, female hygiene, and blood pressure monitoring, there is a need for the development of more point-of-care (POC) diagnosis and prognosis-based biosensors. These advanced biosensors can aid healthcare consultants in diagnosing diseases at an early stage. A highly specific, sensitive, selective, multiprocessing, parallel, and miniature-sized biosensor has the potential to revolutionize current medical practices by rapidly detecting even the smallest targets within a short period of time.

Efforts are underway to develop efficient and smart biosensors, and their ongoing development holds promise for the future. As biosensor technology continues to advance, it is expected to have a significant impact across various medical and healthcare applications, benefiting both healthcare professionals and patients.

Reference

1. Biosensors and their widespread impact on human health; Dinesh Bhatia, Sohini Paul, Tania Acharjee, Shrimanata Sundar Ramachairy, Sensors International 5 (2024) 100257.

Chapter 10
Problems and Perspectives: Catalytic Versus Affinity Biosensing, Sensitivity, Selectivity and Specificity, Reproducibility, Calibration and Uncertainty, Regeneration, Signal Enhancement

Analyzing the published data and the novel methods for monitoring compounds such as drugs, hormones, toxins, small molecules, proteins, and pathogens, it is evident that there is a considerable effort directed toward various detection principles. These include the use of labels and label-free methods, switching and displacement strategies, signal turn-on and turn-off moieties, as well as signal enhancement and amplification techniques. Sensors, whether based on antibodies, aptamers, or phages, are often tailored to operate under specific conditions. In the following section, we will focus our discussion on the challenges associated with sensing platforms.

Enzymes serve as highly effective catalytic biorecognition elements in biosensors due to their specific substrate interactions, which generate measurable products such as protons, electrons, or light. Glucose biosensors, with over 60 years of development, exemplify the successful integration of enzymatic sensing. From early glucose detection to recent adenine biosensors, a wide range of biomarkers have been monitored using innovative catalytic systems. Recent advances have leveraged bionanomaterials—such as nanoparticles, nanowires, carbon nanotubes, and graphene—to enhance the performance of enzyme biosensors, owing to their excellent conductivity and biocompatibility. Examples include ZnS, CdS, Fe_3O_4, AuNPs, hollow gold nanospheres, and multiwall carbon nanotubes. One significant advantage of catalytic biosensors is their potential for real-time, in situ monitoring, making them suitable for wearable or minimally invasive platforms. However, they are limited in their ability to detect macromolecular analytes like proteins, which are not amenable to catalytic conversion. For such targets, affinity-based biosensors are preferred, particularly those employing antibodies. Antibodies bind specifically to antigens via non-covalent interactions, forming highly selective and sensitive Ag–Ab complexes. These complexes are the basis of immunoassays, which are widely used for detecting hormones, enzymes, pathogens, toxins, and tumor markers—typically at concentrations ranging from 10^{-12} to 10^{-9} mol L^{-1}.

Commercial immunoassays rely on the robust formation of Ag–Ab complexes, with detection achieved using a variety of labels such as enzymes (in enzyme-linked

A. Simonian et al., *Biosensors as Analytical Tools for the 21st Century*,
https://doi.org/10.1007/978-3-032-01726-0_10

immunosorbent assays, ELISA), radioisotopes (radioimmunoassays), or chemiluminescent compounds (chemiluminescent immunoassays). Among these, ELISA is the most widely adopted method, available in competitive and sandwich formats. Its popularity stems from its sensitivity, speed, cost-effectiveness, simplicity, and applicability in field diagnostics. The integration of these assays with physical signal transducers has led to the emergence of immunosensors—devices that translate antigen–antibody interactions into measurable electronic or optical signals.

Sensitivity: The required sensitivity for detecting a target analyte depends on its concentration in the detection medium. While electrochemical techniques have achieved significant success, especially with aptamer hairpins that exhibit increased stability and affinity, these strategies can sometimes suffer from low sensitivity. For instance, during the immobilization process, one end of the aptamer hairpin is linked to a redox moiety, while the other end is attached to the electrode surface, and the electrochemical signal is influenced by the proximity of the redox moiety and the conformational changes of the hairpin. This can introduce potential discrepancies in interpreting the signal, which relies on the binding affinity of the target. In particular, signal generation is limited due to distance barriers between the redox species and the electrode surface, leading to false positives. Additionally, the stability of the stem-loop structure when immobilizing aptamer hairpins may be questioned.

To enhance sensitivity, researchers have explored an electrochemical aptamer-based ATP assay using homogeneous solutions, but no attempts have been made to construct aptamer-based biosensors using this approach. While substantial progress has been made using homogeneous solutions, studies have revealed that low-affinity binding constants between the target species and their specific aptamers were obtained, resulting in lower sensitivities.

More recently, techniques such as electrochemical impedance spectroscopy, surface plasmon resonance (SPR), and fluorescence spectroscopy have utilized thrombin aptamers as their target analyte, with much of the research focusing on the sequence in aptamer-based assays, sensors, and biosensors. However, these methods have exhibited drawbacks including low sensitivity, poor detection limits, lengthy measurements, and extended sample preparation times.

To address these limitations and enhance sensitivity, a novel method using Surface-Enhanced Raman Spectroscopy (SERS) in conjunction with magnetic beads as supporting materials has been reported. This method demonstrates high sensitivity for the detection of a lung cancer marker antigen. In comparison to previously reported SERS aptasensors, this approach offers several advantages, including improved sensitivity and a linear response in the lower concentration range, ranging from 0.27 pM to 0.27 nM, with shorter measurement times (less than 1.5 h).

Another innovative approach employed a polymer consisting of poly(m-phenylenediamine) (PMPD) rods as an effective sensing platform to create a novel fluorescent aptasensor. This sensor achieved an exceptionally high sensitivity, boasting a detection limit as low as 100 pM.

Selectivity and Specificity: Numerous affinity-based biosensors employing biorecognition elements such as antibodies, aptamers, and phages have been developed, with thrombin often serving as the target analyte. The rationale behind this choice is that larger targets like thrombin possess a greater number of functional groups and structural ligands that can be captured through hydrogen bonds, electrostatic interactions, or hydrophobic bonds. Consequently, there is a persistent demand for larger targets in biosensing applications. However, the detection of small molecules presents challenges due to their relatively low dissociation constants and poor detection limits.

Less frequently used small molecules in biosensing include cocaine, theophylline, kanamycin, growth factors, tobramycin, tetracycline, and aflatoxin. Detecting these molecules is challenging because simple direct binding procedures for developing aptamer-based assays may not be applicable, as aptamers may struggle to bind directly to these targets. Efforts to improve sensitivity in detecting these small molecules have been reported.

Selectivity issues have prompted the development of multianalyte detection aptasensors capable of detecting proteins, small molecules, and various analytes simultaneously. Recent advancements have showcased bifunctional aptamers capable of binding to two different targets. Additionally, innovative approaches have emerged, such as a sensitive biosensor for adenosine and lysozyme using $[Ru(NH_3)^6]^{3+}$ as a signal transducer and gold nanoparticles (AuNPs) for amplification. However, these sensors are often limited in their ability to detect multiple targets within a single sample, highlighting the need for versatile systems.

More recently, researchers have demonstrated the one-spot detection of adenosine and thrombin using voltammetry, and similar work has been conducted using the ECL technique as a novel strategy. Furthermore, a universal fluorescence aptasensor for detecting proteins, ions, and small molecules has been designed and reported. Assessing selectivity often involves evaluating cross-reactivity, where an aptamer may bind to more than one target analyte due to structural similarities, leading to false positives. Cross-reactivity has been observed in selectivity experiments between OTA and AFM1, both of which share the same molecular weight, using the (Fe_3O_4/PANi) integrated electrode aptasensor. In such situations, Zheng et al. demonstrated the importance of evaluating the sensing system's selectivity by detecting signal changes induced by other mycotoxins.

Specificity and selectivity issues in the detection of pathogenic bacteria should be rigorously evaluated to distinguish real signals from false positives. In one embodiment, specificity was demonstrated through a competitive inhibition assay. For this purpose, different concentrations of phage were preincubated with a known concentration of *Staphylococcus aureus* for 1 h to block the receptor sites from binding. The solutions were then introduced through the sensor. Another phage-based incubation approach used a 360-fold excess of bovine serum albumin in a mixture with galactosidase and found only a 2% reduction in biosensor signal, demonstrating excellent selectivity.

Reproducibility: One of the key challenges associated with aptamer- and phage-based sensors is the reproducibility of results. The occurrence of errors depends on the technique used and the detection range. For instance, SPR-based optical immunosensors are commonly reported to have low detection levels and large error deviations, typically within the range of 2–8%. Consequently, these instruments are less commonly used than amperometric immunosensors. While aptamers have seen considerable success, aptasensors also face similar challenges in terms of repeatability.

For example, eight different samples of OTA with similar concentrations (10 ng/mL) were independently measured using eight different aptasensors, and the relative standard deviation (RSD) was found to be 4.7%, indicating acceptable reproducibility. In other examples, several novel aptasensors developed for the detection of thrombin also unfortunately exhibit acceptable relative standard deviation (RSD). Some of these previous aptasensors are reviewed here. The work by Jiang et al. in thrombin detection showed standard deviation values ranging from 4.4 to 7.3%, with an average reproducibility. Similarly, highly sophisticated G-quadruplex DNAzyme-modified electrodes obtained reproducible variations as high as 7.14%. The RSD of six replicate determinations using a hairpin aptamer was 6.8% at 1 nM/L ($n = 6$).

A novel strategy developed by Yuan et al. demonstrated inter-assay coefficients of variation in triplicate different assays, approximately 4.1% and 5.6% at 0.1 and 10 nM thrombin, respectively. RSD values for other aptasensors using magnetic nanoparticles, splitting DNAzyme, and exonuclease ranged from 2.4 to 4.4% in the detection of thrombin. In phage-based biosensors, reproducibility mainly depends on the stability of the bioreceptor and the fabrication of the biosensor itself. In the work conducted by Tlili et al., the coefficient of variation was assessed through intra- and inter-assay measurements five times with the same concentration of *Escherichia coli* (108 cfu/mL) using three different biosensors. The values obtained were 4.3 and 10.7%, indicating acceptable reproducibility.

In conclusion, despite novel improvements, the problem of uncertainty remains consistently unavoidable and presents a significant challenge. Nevertheless, some aptasensors have reported RSD values of less than 4.3%, offering promise for future technology.

Calibration and Uncertainty: Assessing an ideal sensor's performance relies on a reliable calibration method. Often, much effort in sensor development assumes that calibrations are conducted under in vitro (buffer) conditions. However, these assumptions are not always valid for detecting molecules present in biological fluids under physiological conditions. In fact, this approach becomes complex due to several factors that can significantly affect the concentrations of the target species, leading to considerable variability. One particular issue with electrochemical sensors is the necessity for repetitive calibration for each measurement. This can be mitigated by conducting simultaneous measurements under conditions that closely mimic biological fluids or serum solutions. Although many methods involve using serum solutions as a quick confirmation measurement, the standard calibration curve is often established under different conditions. To minimize measurement uncertainty, it is crucial to use similar medium conditions for both standard calibrations and sample measurements.

Furthermore, there is a desire to develop reliable aptasensors suitable for complex real samples. However, this can introduce challenges when determining concentrations of small molecules, ions, and enzyme activities, as buffers typically contain electrolyte ions (Na^+, K^+) for electrochemistry. Recently, upconversion (UC) nanomaterials have shown good compatibility in biological samples.

Regeneration: Biosensor technology has advanced in the development of smart and unique sensors. Unfortunately, many of these sensors are designed for one-time use only. Therefore, several successful attempts have been made to regenerate sensors for multiple uses. However, even with novel strategies, these sensors not only need to overcome the challenge of regeneration but also maintain good precision and sensitivity after reuse.

For instance, electrochemical aptasensors with surface-confined aptamer probes require significant conformational changes induced by target binding. These complex immobilization procedures often face challenges in sensor regeneration due to the strong binding of the aptamers with the analytes.

In another study, while the sensor exhibited high sensitivity and reproducibility, it failed to achieve satisfactory regeneration due to the separation of thiol groups attached to the gold surface under applied potential. Moreover, sensors do not always maintain their sensitivity after recycling, necessitating the use of multiple concentrations of the bioreceptor in the aptamer layer. To address this limitation, more convenient and general covalent binding approaches are being explored.

Signal Enhancement: Aptasensors developed so far have utilized AuNPs and Au NSp as amplification elements, which have been extended to include strand displacement amplification, rolling circle amplification, and nicking endonuclease signal amplification methods. These methods often rely on protein enzymes for amplification, making them sensitive to reaction conditions such as temperature and pH [1].

Reference

1. Arugula, M. A. & Simonian, A. Novel trends in affinity biosensors: current challenges and perspectives. *Meas Sci Technol* 25, 032001 (2014).

Chapter 11
Laboratory Protocols

11.1 Electrochemistry Lab: Study of Kinetics and Enzyme Electrochemical Reaction Using Cyclic Voltammetry Techniques

Introduction

1. Cyclic voltammetry is the most versatile electroanalytical technique used for investigating the mechanisms of electroactive species. In this laboratory experiment, we utilize the Fe^{III}(CN)63-/Fe^{II}(CN)64 couple as an example of an electrochemically reversible redox system to introduce some fundamental concepts of cyclic voltammetry. Through this experiment, students will become familiar with the basic principles of voltammetry and should be able to determine the following parameters: the formal reduction potential (Eo′); the number of electrons transferred in the redox process (n); the diffusion coefficient (D); electrochemical reversibility; and the effects of varying concentration (C) and scan rate (v).
2. Electrochemical detection of **thiocholine** using acetylcholine esterase enzyme.

$$\text{Acetylthiocholine chloride} + H_2O \rightarrow \text{thiocholine} + \text{acetic acid}$$

Apparatus

All electrochemical measurements will be conducted using a CHI 660 electrochemical analyzer (CH Instruments, Austin, TX) connected to a computer. Cyclic voltammetry measurements were performed using a unicell glassy carbon (GC) electrode with a 3-mm diameter as the working electrode, Ag/AgCl (3 M KCl) as the reference electrode, and platinum wire as the counter electrode. The glassy carbon electrodes were polished using 1, 0.1, and 0.05 μm alumina polish (Buehler) and then rinsed with DI water. Subsequently, the electrodes were sonicated in DI water for 15 min.

A. Simonian et al., *Biosensors as Analytical Tools for the 21st Century*,
https://doi.org/10.1007/978-3-032-01726-0_11

The connections between the potentiostat and the cell are as follows: **black/working electrode, white/reference electrode, and red/auxiliary electrode**.

1. **Experiment**

   (1) A 50 mL stock solution of 50 mM $K_3Fe(CN)_6$ in 0.1 M KNO_3 is prepared by the TA. Dilute this to 2 mM final into 10 mL of PBS buffer in the electrochemical reaction cell.

   (2) Immerse the electrodes and connect all the respective electrodes to the potentiostat. Open the software and select cyclic voltammetry under Technique icon **T** and enter the parameters.

Initial *E*	– 0.2 V	Initial direction	Negative
High *E*	0.6 V	No of segments	2
Low *E*	– 0.2 V	Sensitivity	10 μA/V
Scan rate	0.02 V/s		

   (3) Run a CV. The cyclic voltammogram (CV) will appear on screen as it is generated and will be rescaled automatically upon conclusion. Save the file with appropriate name.

   (4) Select **Parameters** in the **T** menu and change the scan rate from 0.02 to 0.05 V/s. Run the experiment again and save the data. Repeat this procedure for scan rates of 0.1, 0.2, 0.5 and 1 V/s.

   (5) Determine the peak potential (E_{pa}, E_{pc}) and peak current (i_{pa}, i_{pc}). See illustrative figure.

2. **Experiment**
   A 10 mL stock solution of 10 mM Acetylthiocholine (ATCh) in 10 mM PBS (pH 7.4) is prepared, and serial dilutions of this solution will be added to give of 0, 20, 40, 80, 160, 320 μM ATCh in final 10 mM PBS buffer used for electrochemical measurements. A solution of unknown ATCh concentration is provided for analysis. Set the parameters in the parameter menu.

Initial *E*	0.15 V	Initial direction	Negative
High *E*	0.85 V	No of segments	2
Low *E*	0.15 V	Sensitivity	10 μA/V
Scan rate	0.02 V/s		

   (1) In a clean electrochemical cell, add the following volumes of buffer and enzyme solutions starting from No. 1 provided in the data sheet. Mix them well. Immerse the electrodes, connect and then add the substrate (Acetylthiocholine chloride) solution. Wait for the reaction to proceed for 1 min. Then run the CV. Save the data with appropriate name.

(2) Repeat this procedure for each substrate concentrations by adding volumes [μL] provided in the table. Plot current versus concentration (for enzymatic reaction).

Sample	[ATCh] (μM)	Amount of 10 mM ATCh added (μL)	Amount of enzyme (μL)	Amount of PBS buffer (mL)
1	0	0	10	10
2	20	20	10	10
3	40	20	10	10
4	80	40	10	10
5	160	80	10	10
6	320	160	10	10

(3) In a clean electrochemical cell, add the same volumes of buffer and enzyme solutions provided in the data sheet. A sample of unknown concentration will be given for CV measurement. Calculate the unknown concentration based on the plot established in step 2.

Submit

1. Multigraph of scan rates
2. Current versus scan rate plots
3. ΔE versus scan rate plots
4. Current versus concentration plots (for enzymatic reaction)
5. Calculation of unknown concentration (for enzymatic reaction).

Results

1. Multigraph of scan rates

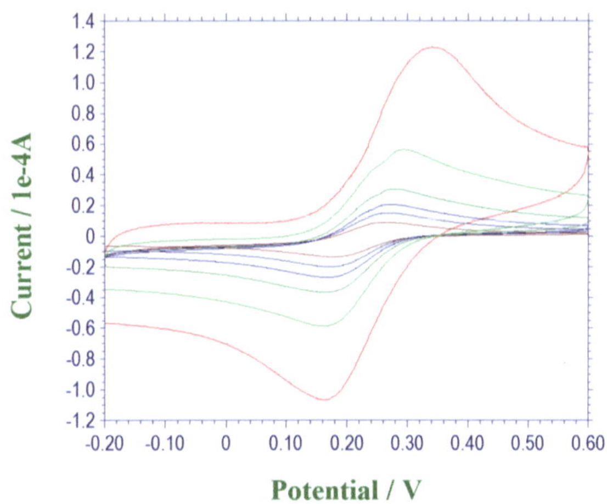

2. Current versus scan rate plots

Scan rate (V/s)	Peak current (μA)
0.02	11.84
0.05	18.33
0.1	24.38
0.2	33.89
0.5	54.29
1	101.9

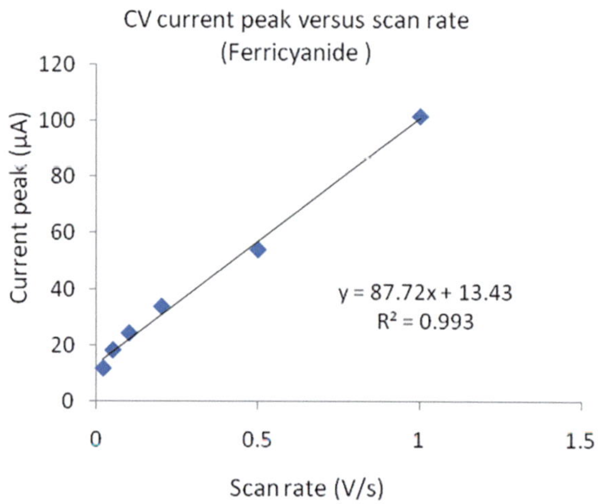

3. ΔE versus scan rate plots

Scan rate (V/s)	E_{pa} (V)	E_{pc} (V)	ΔE
0.02	0.267	0.174	0.093
0.05	0.267	0.174	0.093
0.1	0.275	0.165	0.11
0.2	0.276	0.164	0.112
0.5	0.294	0.162	0.132
1	0.342	0.162	0.18

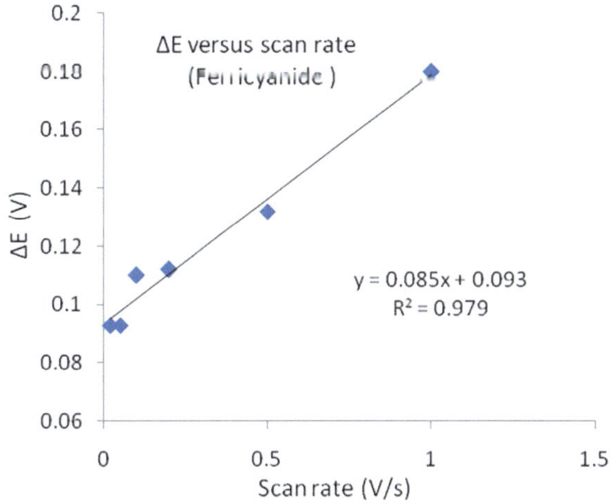

4. Current versus concentration plots (for enzymatic reaction)

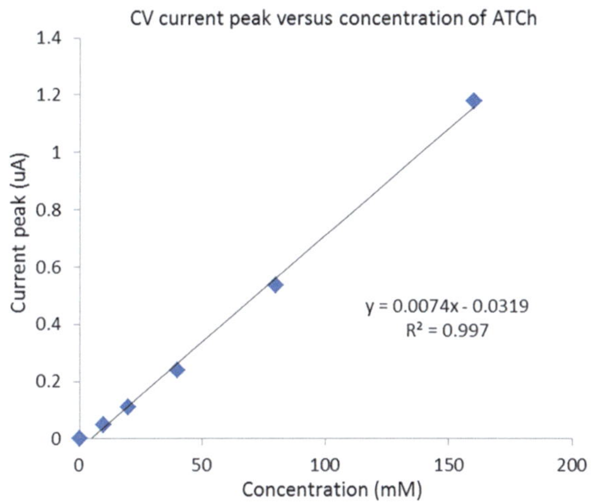

Detection of Tricresyl Phosphate (TCP) Using Amperometric Techniques

(1) Conversion of TCP to cresol

- Preparation of hydrolyzing column
  Prepare slurries of modified alumina by mixing sodium hydroxide and anhydrous aluminum oxide (1:10) in water and stirring for 1 h. Subsequently evaporate the water by placing the slurry in a vacuum chamber. Agitate the dried powders until all the agglomerated lumps are removed and obtain a flowing powder. Pack about 150 mg of this powder into a PVC tube with filters on the either end which constitutes hydrolyzing column.

- Conversion of TCP to cresol
  Transfer solutions containing different concentrations of TCP in methanol into hydrolyzing column and then flush with 10 mM PBS buffer (pH 7.4). Collect all the hydrolyzed products (cresol) for amperometric measurement.

(2) Demonstration of automatically sampling box

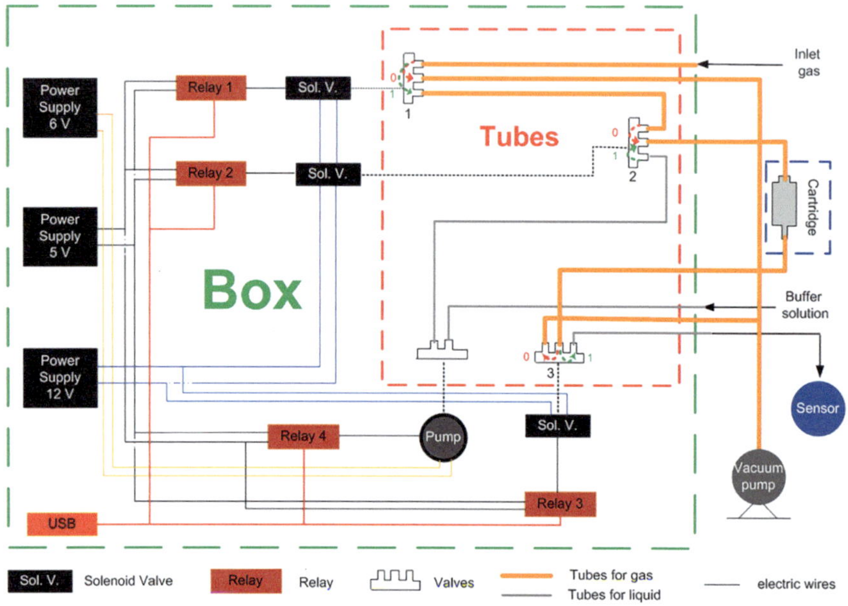

(3) Edot: SDS polymerization of working electrode with cyclic voltammetry

Sample	Edot: SDS
Initial *E* (V)	0
High *E* (V)	1.2
Low *E* (V)	0
Scan rate (V/s)	0.05
Segment	20

(4) Flow-injection amperometric measurement of cresol

Sample	Cresol
Concentration (μM)	1, 5, 25, 50
Potential (V)	0.8
Flow rate (mL/hr)	5.0
Buffer	10 mM PBS buffer (pH 7.4)

(5) Plot the current versus concentration

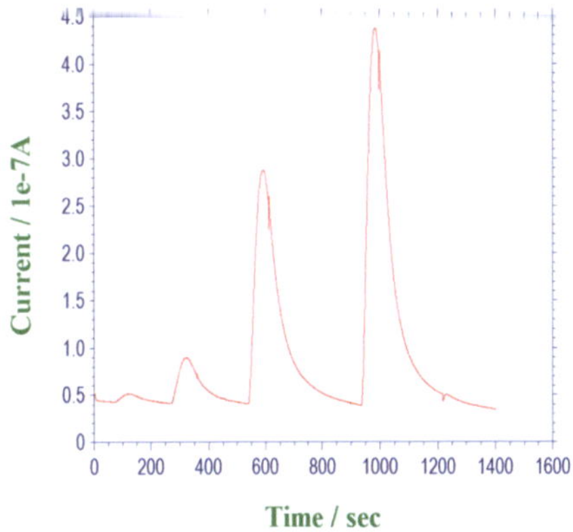

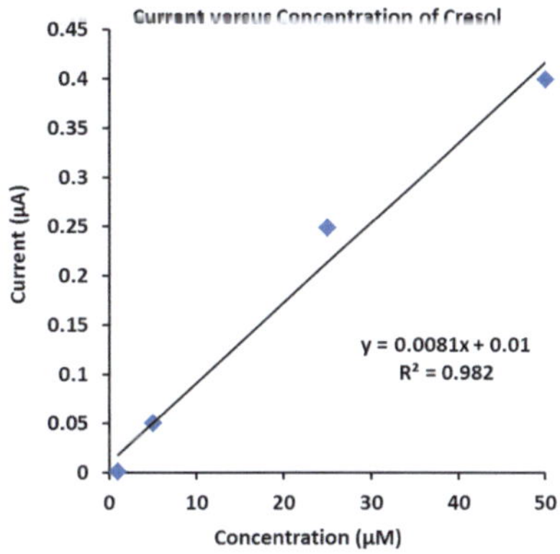

11.2 Spectrophotometric Lab: Hands on Experience Using Common UV–Vis and Nanodrop Instruments, Determination of DNA, and Protein Concentrations, Enzyme Kinetics, Determination of V_{max}, K_M, K_{cat}

Enzymes are protein catalysts, which influence the kinetics but not the thermodynamics of a reaction. They increase the rate of a chemical reaction and do not alter the equilibrium. The study of the rate at which an enzyme works is called enzyme kinetics.

Enzyme Velocity as a Function of Substrate Concentration

If the enzyme velocity at many different concentrations of substrate is measured, the graph generally looks like this:

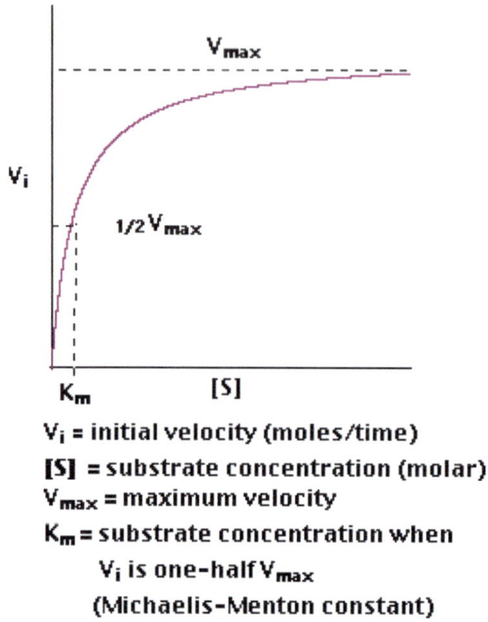

Enzyme velocity as a function of substrate concentration often follows the Michaelis–Menten equation:

$$v_0 = \frac{V_{\max}[S]}{K_M + [S]}.$$

$V_{\max}$ (maximum velocity) is the limiting velocity as substrate concentrations get very large. The maximum observable rate is $V_{\max}$ and the data will asymptote to this value at high $[S]$. At low $[S]$, the reaction rate is generally linearly proportional to the $[S]$. $V_{\max}$ (and V) are expressed as units of moles of product formed per second.

K_M (Michaelis constant) is the concentration of substrate that leads to half-maximal velocity. K_M is expressed in units of concentration, Molar units. A low value of K_M is interpreted as a high affinity of the enzyme for the substrate, and a large value for K_M is interpreted as a weak affinity of the enzyme for the substrate. $V_{\max}$ and K_M are intrinsic properties of the particular enzyme/substrate combination.

k_{cat} (turnover number) represents the maximum number of molecules of substrate which can be converted into product per enzyme molecule per unit time (which only happens if the enzyme is "saturated" with substrate) and thus is often known as the turnover number.

Equipment

1. Spectrophotometer
2. Cuvettes.

Reagents

1. Glucose-6-phosphate dehydrogenase (enzyme)
2. NAD+ (co factor)
3. Phosphate Buffer 100 mM pH 7.4
4. Glucose-6-phosphate (substrate).

Procedure

Step 1: Measurement of the Oxidation of Substrate by Enzyme G-6-PDH

1. In the cuvette add the solutions as per the table and read the product @ 340 nm (NADH) and monitor for 1 min.
2. Record the slope from the graph.
3. Calculate the initial velocities for the recorded slopes.

$$V_0 = \frac{\Delta \text{Abs} \cdot \text{min}^{-1}}{\varepsilon} \cdot \frac{1}{60\,\text{s}} \cdot 1\,\text{cm} \cdot \frac{1\,\text{L}}{1000\,\text{mL}} \cdot \frac{10^6\,\mu\text{mol}}{\text{mol}}$$

The extinction coefficient (ε), will be different depending on the product that is monitored. For NADH monitoring, the extinction coefficient is 6220 M^{-1} cm^{-1} @ 340 nm.

4. Copy the slope into Excel sheet and calculate initial velocities. Then in ORIGIN plot V_0 versus [substrate] and fit to the Michaelis equation to determine V_{max} and K_M.

$$v_0 = \frac{V_{max} \cdot [S]}{K_M + [S]}.$$

In ORIGIN software use analysis, sigmoidal fitting, function, and Michaelis–Menton. The software automatically displays the values of V_{max} and K_M.

5. Then calculate k_{cat}:

$$k_{cat} = \frac{V_{max}}{[E_T]},$$

where $[E_T]$ = total enzyme concentration (number of micromoles of enzyme) in the reaction cuvette. First calculate number of moles in volume of enzyme used from the stock solution. Then use molecular weight of the enzyme for molar concentration calculation.

6. Divide the k_{cat} by the K_M to calculate the efficiency of the enzyme.

$$\text{Efficiency of enzyme} = k_{cat}/K_M$$

[G-6-P] (mM)	Amount of 20 mM G-6-P (μL)	Amount of 10 mM NAD+ (μL)	Amount of enzyme (μL)	Amount of phosphate buffer (μL)	Slope from graph
0	0	15	50	935	
0.1	5	15	50	930	
0.25	12.5	15	50	925	
0.5	25	15	50	910	
0.75	37.5	15	50	895	
1	50	15	50	885	
1.25	62	15	50	860	
1.5	75	15	50	870	

Experimental Results

[G-6-P] (mM)	Amount of 20 mM G-6-P (μL)	Amount of 10 mM NAD+ (μL)	Amount of enzyme (μL)	Amount of phosphate buffer (μL)	Slope from graph
0	0	15	50	935	0.0002
0.05	2.5	15	50	933	0.0973
0.1	5	15	50	930	0.1456
0.25	12.5	15	50	922	0.2578
0.5	25	15	50	910	0.3546
0.75	37.5	15	50	897	0.4060
1	50	15	50	885	0.4119

Convert the Slope to Velocities (See Step 3)—Plot Velocity Versus [G-6-P]

[PX] (mM)	Slope from graph	Initial velocity
0	0.0002	5.35906E−06
0.05	0.0973	0.002607181
0.1	0.1456	0.003901393
0.25	0.2578	0.006907824
0.5	0.3546	0.009501608
0.75	0.4060	0.010878885
1	0.4119	0.011036977

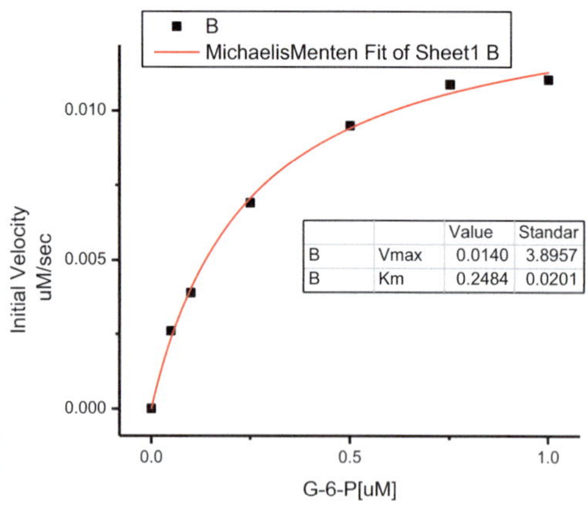

From graph, $V_{\max} = 0.0140$ μmol, $K_M = 0.248$ mM

$$k_{cat} = \frac{V_{max}}{[E_T]}$$

$K_{cat} = 0.0140$ μmol/s/13.65 × 10^{-7} μmol = 10,240 s^{-1}
Efficiency of enzyme = k_{cat}/K_M = 10,240 s^{-1}/0.000248 M = 0.413 × 10^8 M^{-1} s^{-1}
To calculate $[E_T]$
[WT] in M
= 3.5 mg/ml/(128,000 g/mol * 1000 dilution)
= 2.73 × 10^{-8} mol/L or M
50 μL of this enzyme was used.
= 2.73 × 10^{-8} mol/L * 50 μL * 1 L/10^6 μL
= 13.65 × 10^{-13} mol
= 13.65 × 10^{-7} μmol.

11.3 Surface Plasmon Resonance Lab: Calibration of SPR Sensor, SPR Data Handling

When choosing the appropriate sensor chip and immobilization method, several factors should be considered, such as the nature of the ligand and analyte, stability of the interaction, sensitivity requirements, and the desired assay format. It is also important to optimize the immobilization conditions, including the pH of the immobilization buffer, ligand concentration, and immobilization time (Table 11.1).

The immobilization can be performed using different strategies depending on the sensor surface. For gold surfaces, self-assembled monolayers (SAMs) formed through thiol chemistry are commonly used. Dextran surfaces can be immobilized through covalent coupling methods or amine coupling using EDC/NHS chemistry. Carboxymethylated dextran surfaces also utilize amine coupling or EDC/NHS chemistry. Nitrocellulose surfaces allow passive adsorption or covalent coupling. Silicon

Table 11.1 Types of sensor surfaces and methods for immobilization

Sensor surface types	Immobilization methods
Gold surface	Self-assembled monolayers (SAMs), thiol chemistry
Dextran surface	Covalent coupling, amine coupling, EDC/NHS chemistry
Carboxymethylated dextran surface	Amine coupling, EDC/NHS chemistry
Nitrocellulose surface	Passive adsorption, covalent coupling
Silicon surface	Silane chemistry, covalent coupling
Polystyrene surface	Passive adsorption, covalent coupling
Glass surface	Silane chemistry, covalent coupling
Polymer surface	Covalent coupling, EDC/NHS chemistry

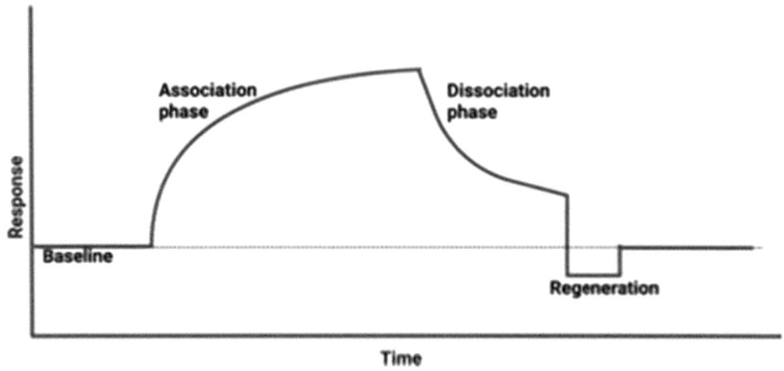

Fig. 11.1 Interaction analysis sensorgram phases

surfaces are typically functionalized using silane chemistry or covalent coupling. Polystyrene surfaces can be immobilized through passive adsorption or covalent coupling. Glass surfaces can be functionalized using silane chemistry or covalent coupling. Polymer surfaces can be immobilized using covalent coupling or EDC/NHS chemistry.

The choice of sensor surface and immobilization method should be carefully evaluated based on the specific requirements of the experiment to ensure optimal ligand immobilization and subsequent analyte interaction measurements in the surface plasmon resonance (SPR) experiment (Fig. 11.1).

The sensorgram encompasses the following steps:

Baseline: The baseline phase is the initial phase of the sensorgram where the analyte-free buffer is flowed over the immobilized ligand on the sensor surface. This establishes a stable baseline signal, which is important for accurate measurement of the subsequent association and dissociation phases.

Association: During the association phase, the analyte solution is introduced and allowed to interact with the immobilized ligand on the sensor surface. The binding of the analyte to the ligand results in an increase in the signal, which represents the formation of the ligand-analyte complex. The association phase provides information about the rate and extent of binding between the ligand and analyte.

Steady State: If reached, the steady-state phase occurs when the binding between the ligand and analyte reaches equilibrium. At this point, the sensorgram signal plateaus, indicating that the association and dissociation rates have balanced out. The duration of the steady-state phase can vary depending on the kinetics of the interaction and the concentration of the analyte.

Dissociation: Following the association phase, the analyte flow is stopped, and the dissociation phase begins. During this phase, the analyte is removed from the sensor surface, and the ligand-analyte complex dissociates. The dissociation phase provides information about the stability of the ligand-analyte interaction and the dissociation rate.

Regeneration: After the dissociation phase, a regeneration step may be performed to remove any remaining bound analyte and regenerate the sensor surface for subsequent measurements. Regeneration is typically achieved by flowing a regenerating solution over the sensor surface, which disrupts the ligand-analyte interaction and removes any residual analyte. The regeneration step is important for ensuring accurate and repeatable measurements in subsequent cycles of the experiment.

Flow Buffer: The flow buffer is a solution used to condition the sensor chip and check for any instability or drift in the baseline signal. It ensures that the sensor chip is in an optimal state for the interaction analysis and helps identify any issues with the system. Injection spikes in the baseline signal indicate problems with the system and may require troubleshooting and cleaning before proceeding with the experiment.

It is crucial to maintain a stable baseline and carefully monitor the sensorgram throughout the experiment to ensure reliable and accurate measurement of the ligand-analyte interaction.

Studying Lipid–Protein Interactions with the L1 Sensor Chip

The L1 sensor chip is specifically designed for studying lipid–protein interactions using surface plasmon resonance (SPR), as shown in Fig. 11.2. It has lipophilic groups incorporated into the dextran matrix, allowing for the capture of intact lipid vesicles or the formation of a lipid bilayer on the chip surface. This chip provides a biocompatible environment for studying the binding affinity and specificity of lipid–protein interactions.

To prepare the L1 chip for experimentation, the following steps are typically followed:

1. Wash the chip surface: The chip surface is washed with solutions such as 20 mM 3-[(3-cholamidopropyl) dimethylammonium]-1-propanesulfonate (CHAPS) and 40 mM octyl glucoside. This step helps in cleaning the chip surface and removing any contaminants.
2. Coat the chip with liposomes: Liposomes, which mimic the lipid composition of the biological membrane being studied, are injected onto the chip surface at a controlled flow rate. This step allows the immobilization of lipids on the chip and creates a surface that resembles the biological membrane.

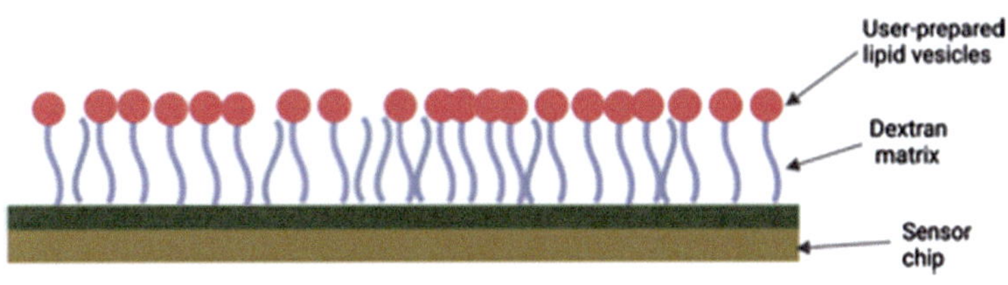

Fig. 11.2 L1 sensor chip for lipid–protein interactions

3. Stabilize the lipid surface: After coating the chip with liposomes, a stabilization step is performed using a flow of 50 mM sodium hydroxide (NaOH) solution. This helps in removing loosely bound lipid vesicles from the chip surface.
4. Block the exposed chip surface: To prevent non-specific binding, the exposed chip surface is blocked by injecting fatty acid-free bovine serum albumin (BSA) onto the chip. This step reduces non-specific interactions and ensures that only specific lipid–protein interactions are measured.

Once the chip surface is prepared, the interaction analysis can be performed by injecting the protein of interest at different concentrations onto the chip. The binding response, represented by the response units (RU), is measured as the protein interacts with the immobilized lipids. The interaction can be monitored in real-time, and the steady-state RU values at each protein concentration can be used to calculate the apparent dissociation constant (K_d) using the equation $R_{eq} = R_{max}/(1 + K_d/C)$, where R_{eq} is the binding signal obtained at steady state for each concentration (C) and R_{max} is the binding signal at saturation of the analyte-ligand system.

It is important to troubleshoot any experimental problems that may arise during the study. Table 11.2 summarizes common problems encountered in SPR experiments and provides possible solutions for each issue.

By carefully preparing the L1 chip surface and conducting the experiment using appropriate protein concentrations and conditions, valuable information about lipid–protein interactions can be obtained. The L1 chip provides a reliable platform for studying the affinity and specificity of these interactions, contributing to our understanding of the roles of lipids and proteins in cellular processes.

Table 11.2 Troubleshooting common problems in SPR experiments

Problem	Possible solution
Drift in SPR curve	Check for temperature stability, stabilize the system
Jumps in SPR curve	Check for air bubbles, stabilize the system
Spikes in SPR curve	Check for contamination, clean the system
Irreproducible data	Verify experimental conditions, optimize protocol
High non-specific binding	Optimize blocking step, adjust surface conditions
Weak or no binding signal	Check protein quality, optimize protein concentration, pH, buffer
High variability in replicate runs	Optimize experimental conditions, increase replicates

11.4 Fluorescence Lab

The experiment involves analyzing the concentration of vitamin B2 (riboflavin) using its native fluorescence. Riboflavin is a water-soluble vitamin found in various foods and exhibits strong fluorescence when exposed to light. It decomposes under irradiation into two highly fluorescent products called lumichrome and lumiflavin.

To perform the analysis, it is crucial to work with riboflavin samples at concentrations below the part per million (ppm) level. Therefore, extreme care must be taken when cleaning and handling glassware and solutions throughout the experiment.

During the first lab period, you need to prepare the standard and unknown solutions as described below. Additionally, consult the Teaching Assistant (T.A.) regarding the operation of the UV–Vis spectrophotometer and fluorescence spectrometer. Acquire the absorption spectra of riboflavin and conduct a practice run on the fluorescence spectrometer.

During the second lab period, you will complete the experiment and Part II of the unknown analysis.

Stock Solutions

Here are the instructions for preparing the stock solutions:

1. Obtain the 100 ppm stock standard solution of riboflavin from the refrigerator or T.A.
2. Allow the solution to reach room temperature by leaving it on the counter for a few minutes to ensure consistent volume measurements.
3. Transfer 1 mL of the stock solution to a 10 mL volumetric flask and dilute it using deionized water. This will give you a 10 ppm solution.
4. Using a volumetric pipet, deliver 1.00 mL of the 10 ppm solution to a 100 mL volumetric flask and dilute it with deionized water. This will yield a 100 part per billion (ppb) stock solution.
5. Fill Table 11.3 (not provided) with the appropriate amounts of the 100 ppb riboflavin solution to prepare the remaining standards. Double-check the table with your TA before proceeding.
6. Prepare solutions 2–6 from the table and dilute them with deionized water. Ensure that all glassware is clean and rinsed with deionized water before use.

Table 11.3 Standard riboflavin solutions

Solution #	mL of 100 ppb sol.	Dilute to (mL)	Rib. conc. (ppb)
2		50	2
3		50	10
4		50	20
5		50	40
6		50	50

Note: The content of Table 11.3, which specifies the amounts of riboflavin solution for each standard, is not provided. Please refer to the original experiment instructions or consult your TA for the correct values.

Remember to follow proper lab procedures and safety guidelines while handling chemicals and equipment throughout the experiment.

Unknown Samples

To prepare the milk sample for analysis, follow the procedure outlined below:

1. Locate the dry milk sample on the shelves above the weigh station.
2. Reconstitute the dry milk by following the instructions provided on the sample jar.
3. Take 25 mL of the reconstituted milk sample.
4. Prepare a solution by mixing equal parts of 3 M acetic acid (HAc) and 3 M sodium chloride (NaCl). You will need a total volume of 75 mL of this solution.
5. Using a 25 mL pipet and a 250 or 400 mL beaker, add exactly 75 mL of the HAc-NaCl solution to the 25 mL milk sample. Avoid using a volumetric flask as it may be difficult to stir the solution properly.
6. Stir the mixture for 20 min to ensure thorough mixing.
7. Filter the solution using a Buchner funnel and a vacuum line. Ensure that there is a trap between the filtering flask and the house vacuum line to prevent any contamination.
8. Take a 5 mL aliquot (a measured portion) of the filtered solution.
9. Dilute the 5 mL aliquot to a total volume of 100 mL using deionized water in a volumetric flask. This diluted solution will be used for your analytical measurements.

Remember to handle the milk sample and solutions with care, following proper lab procedures and safety guidelines.

Instrumental

In this experiment, you will be using the Agilent HP8453 UV–Vis spectrometer and the Varian Eclipse Fluorimeter to analyze riboflavin. The procedure is divided into three parts: Approximate excitation wavelength, actual excitation and emission maxima, and standard curve. Here are the steps involved in each part:

Part 1: Approximate Excitation Wavelength

1. Locate a pack of cuvettes near the Agilent HP8453 UV–Vis spectrometer. If you cannot find one, ask your TA for a new pack.
2. Prepare a 10 ppm solution of riboflavin by diluting the stock solution.
3. Set the spectrometer to acquire an absorption spectrum of riboflavin from 280 to 600 nm. Use distilled water as the reference solution.
4. Print the absorbance spectrum obtained.

Part 2: Actual Excitation and Emission Maxima

1. Using the approximate excitation wavelengths determined in Part 1, obtain two emission spectra of the 50 ppb standard solution. The first spectrum should be collected from the first excitation wavelength (~ 370–700 nm), and the second spectrum from the second excitation wavelength (~ 440–700 nm).
2. Collect two excitation spectra: one from the 50 ppb standard and one from the milk sample you prepared. The wavelength range for both excitation spectra should be from 280 to about 30 nm below the emission wavelength determined in Part 2.
3. Print all the spectra obtained in this section.

Using the Varian Eclipse Fluorimeter

1. Ask your TA for the necessary cuvette for the instrument.
2. Power up the Varian Eclipse Fluorimeter and log into the "Fluorimeter" program.
3. Double click on the Scan icon on the desktop, then click on the Setup button.
4. Set the excitation wavelength to one of the absorption maxima obtained from the UV–Vis spectrum.
5. Set the scan limits for the emission spectrum, starting 30 nm above the excitation wavelength and up to 700 nm.
6. Set the scan speed to Slow (120 nm/min) and both slits to 10 nm.
7. Click on the Reports tab and set the Threshold value to 5. Click OK.
8. Start the scan by clicking on the green traffic light icon. Name your sample when prompted and click OK.
9. Print the spectrum when the scan is complete.
10. Repeat the above steps for the excitation spectrum, but select the excitation box and record the spectrum from 280 to about 30 nm below the emission wavelength.
11. Repeat the excitation spectrum collection for the milk sample.

Part 3: Standard Curve

1. Double click on the Simple Reads icon on the desktop.
2. Click on the Setup button and set the emission and excitation wavelengths.
3. Set the slits to 10 nm and the Average Time to 5 s.
4. Click on Read to start data acquisition.
5. Change the excitation wavelength and repeat the procedure for all your standards.
6. Repeat the measurement three times for the unknown sample.
7. Print the data files obtained.
8. Create a standard curve in Excel or Google Sheets using the data obtained.
9. Answer the post-lab questions, including the inclusion of all spectra and the calibration curve.

Remember to follow proper lab procedures and safety guidelines throughout the experiment.

References

1. Catalytic Hydrolysis of Tricresyl Phosphate by Ruthenium (III) Hydroxide and Iron (III) Hydroxide towards Sensing Application", Lang Zhou, Bryan Chin and Alex L. Simonian, Sensors (Basel), 2020 Apr 18;20(8):2317. https://doi.org/10.3390/s20082317.

Index

A. Simonian et al., *Biosensors as Analytical Tools for the 21st Century*, https://doi.org/10.1007/978-3-032-01726-0

MIX
Papier aus verantwortungsvollen Quellen
Paper from responsible sources
FSC® C105338

If you have any concerns about our products,
you can contact us on
ProductSafety@springernature.com

In case Publisher is established outside the EU,
the EU authorized representative is:
Springer Nature Customer Service Center GmbH
Europaplatz 3, 69115 Heidelberg, Germany

Printed by Libri Plureos GmbH
in Hamburg, Germany